TOXIC SUBSTANCE STORAGE TANK CONTAINMENT

TOXIC SUBSTANCE STORAGE TANK CONTAINMENT

by

Ecology and Environment, Inc.
Buffalo, New York

and

Whitman, Requardt and Associates
Baltimore, Maryland

np

NOYES PUBLICATIONS
Park Ridge, New Jersey, U.S.A.

Library of Congress Catalog Card Number 84-22697
ISBN: 0-8155-1018-7
ISSN: 0090-516X
Printed in the United States

Published in the United States of America by
Noyes Publications
Mill Road, Park Ridge, New Jersey 07656

10 9 8 7 6 5 4 3 2 1

Library of Congress Cataloging in Publication Data

Ecology and Environment, Inc.
Toxic substance storage tank containment.

(Pollution technology review, ISSN 0090-516X ; no. 116)
Includes bibliographies and index.
1. Hazardous substances--United States--Storage--Handbooks, manuals, etc. 2. Storage tanks--United States--Handbooks, manuals, etc. 3. Waste spills--United States--Handbooks, manuals, etc. I. Whitman, Requardt, and Associates. II. Title. III. Series.
TD811.5.E36 1985 628.5'46 84-22697
ISBN 0-8155-1018-7

Foreword

This book provides basic guidelines upon which a toxic substance storage tank containment assurance and safety program might be based. It includes guidelines for maintenance, inspection, and emergency procedures, as well as references to the appropriate standards and codes with which storage tanks should be in compliance.

Accidental spills or releases of hazardous substances can result in contamination of groundwater and surface water, exposure of populations to toxic or carcinogenic chemicals, destruction of property, severe financial liabilities, and adverse corporate publicity. Therefore, rigorous requirements are necessary for the management and control of hazardous substances.

In order to reduce the occurrence of toxic substance releases from storage facilities, standards for design, maintenance, and operation of new facilities should be evaluated to determine if they are adequate. Older facilities should also be examined. If they do not meet current standards, they should be secured by the implementation of appropriate corrective measures.

This book provides the basis for facility evaluation. It is presented in two parts which cover guidelines and procedures, and technical code application.

For the sake of brevity, throughout the book the following abbreviations or acronyms are used:

EPA	The U.S. Environmental Protection Agency
DHMH	Maryland Department of Health & Mental Hygiene
OEP	The Office of Environmental Programs (DHMH)
SHAG	Science and Health Advisory Group (OEP)
E & E	Ecology and Environment, Inc.
MOSH	Maryland Occupational Safety & Health Administration
DNR	Maryland Department of Natural Resources

The information in the book is from:

> *Toxic Substance Storage Tank Containment Assurance and Safety Program: Guide and Procedures Manual,* prepared by Ecology and Environment, Inc. and Whitman, Requardt and Associates for the State of Maryland, Department of Health and Mental Hygiene, under a grant from the U.S. Environmental Protection Agency, September 1983.
>
> *Toxic Substance Storage Tank Containment Assurance and Safety Program: Guidelines for Application of Technical Codes for Hazardous Materials Storage Tanks,* prepared by Ecology and Environment, Inc. and Whitman, Requardt and Associates for the State of Maryland, Department of Health and Mental Hygiene, under a grant from the U.S. Environmental Protection Agency, September 1983.

The table of contents is organized in such a way as to serve as a subject index and provides easy access to the information contained in the book.

Advanced composition and production methods developed by Noyes Publications are employed to bring this durably bound book to you in a minimum of time. Special techniques are used to close the gap between "manuscript" and "completed book." In order to keep the price of the book to a reasonable level, it has been partially reproduced by photo-offset directly from the original reports and the cost saving passed on to the reader. Due to this method of publishing, certain portions of the book may be less legible than desired.

NOTICE

The material in this book has been reviewed by the State of Maryland Department of Health and Mental Hygiene, Office of Environmental Programs, and approved for publication. Approval does not signify that the contents necessarily reflect the views and policies of the Department of Health and Mental Hygiene, or the United States Environmental Protection Agency, or the publisher, nor does mention of trade names or commercial products constitute endorsement or recommendation for use.

Because hazardous materials vary widely in their characteristics and in the manner in which they should be stored, the material contained within this book can serve only as a guide. It is the responsibility of the storage facility owner to seek the assistance of appropriately qualified professionals with the necessary skills to design a storage system which can be used safely, and which provides the necessary measures for public and environmental protection.

Acknowledgments

Development of this document was accomplished with the assistance of an advisory committee representing the following State of Maryland agencies:

- Department of Health and Mental Hygiene
  - Office of Environmental Programs
    - Air Management Administration
    - Waste Management Administration
    - Science and Health Advisory Group
- Department of Public Safety and Correctional Services
  - State Fire Marshall's Office
- Department of Natural Resources
  - Water Resources Administration
- Department of Licensing and Regulation
  - Maryland Occupational Safety and Health Administration
- Maryland Casualty Company, Baltimore

We also wish to express our appreciation for the guidance and direction provided throughout this project by Mr. K.K. Wu, Toxics Integration Coordinator, U.S. Environmental Protection Agency, Region III, Philadelphia, Pennsylvania. Mr. Wu was always available for our questions and concerns, and proved a willing ear when problems arose.

We also wish to acknowledge the expertise and tireless dedication of the Project's Advisory Committee, who made the actualization of the finished project, in large part, possible. These persons were:

Andrew Alcarese, MOSH
Donald Andrew, Air Management, OEP

John J. Barranger, Maryland Casualty Company
John Bender, Office of the State Fire Marshal
William Burgess, Water Resources Administration, DNR
David Healy, Waste Management Administration, OEP
Frank Henderson, Waste Management Administration, OEP
John E. McQuade, Jr., SHAG, OEP
Othniel Thompson, SHAG, OEP, Chairman
Edwin C. Weber, Water Resources Administration, DNR
Carl York, Air Management Administration, OEP

Acknowledgment should also be made to Max Eisenberg, Ph.D., Director of Science and Environmental Health, OEP, Director of SHAG, and also a member of the Governor's Council on Toxic Substances which provided advice and encouragement.

Contents and Subject Index

PART II
APPLICATION OF TECHNICAL CODES

Part I
Guide and Procedures Manual

The information in Part I is from *Toxic Substance Storage Tank Containment Assurance and Safety Program: Guide and Procedures Manual,* prepared by Ecology and Environment, Inc. and Whitman, Requardt and Associates for the State of Maryland, Department of Health and Mental Hygiene, under a grant from the U.S. Environmental Protection Agency, September 1983.

1. Introduction

1.1 BACKGROUND

Hazardous substances are those chemical and petroleum products which exhibit characteristics of toxicity, ignitability, flammability, reactivity, and corrosivity. Depending on their quantities, concentrations, and physical and chemical characteristics, these substances may pose substantial present or future hazards to human health or the environment if improperly treated, stored, transported, or disposed of. Accidental spills or releases of hazardous substances can result in contamination of groundwater and surface water, exposure of populations to toxic or carcinogenic chemicals, destruction of property, severe financial liabilities, and adverse corporate publicity. Therefore, rigorous requirements are necessary for the management and control of hazardous substances.

The United States Coast Guard (USCG) reported in recent years that up to 27% of all reported spill incidents in the United States, including spills of oil and hazardous substances, occurred from facilities not related to transportation. This amounts to more than 3,200 incidents a year. The majority of all spill incidents are caused by equipment failures (40%) or human error (18%). Many of these spills could have been prevented by the appropriate application of maintenance, testing, and inspection procedures.

Structural failures of storage tanks, resulting in releases of toxic substances to the environment, occur because of inadequate design, or because of improper or infrequent maintenance of the tanks, valves, or transfer lines. Many accidental releases of substances occur during transfer operations, and adequate secondary containment measures often are not provided. Once released to the immediate environment, chemicals may be transported through surface water systems or may leach into groundwater where control and recovery are more difficult. Airborne vapor clouds also may present significant problems.

1.2 OBJECTIVES OF CONTAINMENT ASSURANCE PROGRAM

In order to reduce the occurrence of toxic substance releases from storage facilities, standards for design, maintenance, and operation of new facilities should be evaluated to determine if they are adequate. Older facilities should also be examined. If they do not meet current standards, they should be secured by the implementation of appropriate corrective measures. These measures constitute a

Hazardous Substance Containment Assurance and Safety Program. The objectives of such a program should include:

- Utilization of appropriate criteria for storage tank design and maintenance, based on the most recent chemical, technical, and structural standards;

- Standardization of preventive maintenance and inspection schedules for hazardous substance storage tanks;

- Training of management, maintenance, and inspection personnel in sound practices for hazardous substance control; and

- Providing guidelines for developing a hazardous substance spill prevention program, including recommendations for emergency action and secondary containment.

This Guide and Procedures Manual will provide the basic guidelines upon which a containment assurance and safety program may be based. These include guidelines for maintenance, inspection, and emergency procedures, as well as references to the appropriate standards and codes with which storage tanks should be in compliance.

1.3 SCOPE AND APPLICATION

This manual is intended to provide the basic information needed to reduce the likelihood of a hazardous materials storage system failure. It provides information in the form of guidelines for chemical compatibility, tank design and installation, corrosion control, maintenance and inspection, personnel safety and training, and spill prevention and contingency planning. Although the manual primarily addresses potential problems with hazardous liquids, issues concerning gases and vapors are also discussed. Solid materials, as a class, are not addressed in this manual. Because it is intended to provide guidelines, the manual's scope is necessarily limited to discussion of general applications. Wherever more detailed information may be required for specific circumstances, the user is referred to the most appropriate reference sources. These sources often include technical standards and codes. Because the codes are updated frequently, the user should have the most recent edition. In those cases where a regulatory code cites a technical code by date, the technical code referred to should not be older than the one referenced in the regulation.

Section 2 of the Guide and Procedures Manual identifies the major chemical classes and provides a summary matrix of reactions that may occur among them. The section also provides a comprehensive listing of compatibility of specific chemicals and the major materials used in construction of storage tanks and appurtenances. These data will enable inspection personnel to more readily identify undesirable storage practices and institute appropriate mitigative measures. Section 3 presents design and installation considerations. These include guidelines for tank selection, ventilation, flammability protection, and spill control. Because corrosion is a major problem with regard to storage tanks, corrosion control guidelines are discussed separately in Section 4. Tank maintenance and inspection guidelines are given in Section 5. These include discussions of testing and inspection procedures for specific types of tanks and appurtenances, recommendations on inspection frequency, and criteria for determining the need for corrective action. Safety assurance in toxic substance storage systems is also dependent upon knowledgeable operators. Section 6 of the manual presents elements of a personnel training program covering the safe operation of hazardous material storage systems. The section contains guidelines for safety precautions to be exercised by storage site personnel, selection of protective and monitoring equipment, and a training schedule to meet occupational health and safety standards. To further reduce the possibility of an accidental spill, Section 7 gives the basic elements of Spill Prevention, Control, and Countermeasure (SPCC) plans. The SPCC plans are designed to:

- Insure rapid and accurate detection of emergency situations;
- Provide methods and procedures to minimize environmental impacts;
- Provide methods and procedures to facilitate efficient recovery and removal of spilled material; and
- Provide safety measures for response personnel.

This Guide and Procedures Manual is intended to provide a mechanism for the development of hazardous substance containment assurance programs by industrial managers, city planners, and permitting agencies. It is intended to be a practical tool for both private industry and public regulatory agencies in establishing workable standards and guidelines for toxic substance storage. Its implementation will help solve the long-term problems associated with the containment of hazardous materials.

2. Chemical Compatibility

2.1 INTRODUCTION

A critical issue in the safe handling of hazardous chemicals is their compatibility with other chemicals or materials with which they may come in contact. The combination of two or more incompatible chemicals may have one or more of the following consequences:

- Heat generation,
- Fire,
- Explosion,
- Innocuous gas generation,
- Toxic gas generation,
- Flammable gas generation,
- Uncontrolled polymerization, or
- Solubilization of toxic substances.

In addition to the possible consequences listed above, many construction materials may undergo corrosion, loss of structural integrity, or total destruction when in contact with certain chemicals or a combination of chemicals. Therefore, care must be exercised to avoid the inadvertent mixture of incompatible chemicals and materials. This includes such measures as properly cleaning a previously used tank before filling it with a different chemical, selecting tank and appurtenance materials that are resistant to the chemicals stored, and avoiding storage of incompatible chemicals in proximity with one another.

2.2 CHEMICAL COMPATIBILITY MATRIX

A chemical compatibility matrix has been devised to provide a means for determining the likely consequences of combining chemicals from two classes. Classes of chemicals are listed in Table 2-1. Each class consists of chemical compounds of similar molecular structure (classes 1-31) or similar reactivity characteristics (classes 32-38). An extensive listing of chemical compounds by chemical class is provided in Appendix A.

One would expect that chemicals of similar structure would be of the same class and would undergo similar chemical compatibility reactions. Therefore, the class of a chemical not identified in Appendix A can be determined by locating a listed compound of similar molecular structure. For example, the chemical class of diethyl phthalate can be determined by locating on the list dimethyl phthalate, which has a

Table 2-1

LIST OF CHEMICAL CLASSES

Chemical Class Number	Class Name
1	Acids, mineral, non-oxidizing
2	Acids, mineral, oxidizing
3	Acids, organic
4	Alcohols and glycols
5	Aldehydes
6	Amides
7	Amines, aliphatic and aromatic
8	Azo compounds, diazo compounds, and hydrazines
9	Carbamates
10	Caustics
11	Cyanides
12	Dithiocarbamates
13	Esters
14	Ethers
15	Fluorides, inorganic
16	Hydrocarbons, aromatic
17	Halogenated organics
18	Isocyanates
19	Ketones
20	Mercaptans and other organic sulfides
21	Metal compounds, inorganic
22	Nitrides
23	Nitrites
24	Nitro compounds
25	Hydrocarbons, aliphatic, unsaturated
26	Hydrocarbons, aliphatic, saturated
27	Peroxides and hydroperoxides, organic
28	Phenols and cresols
29	Organophosphates, phosphothioates, and phosphodithioates
30	Sulfides, inorganic
31	Epoxides
32	Combustible and flammable materials
33	Explosives
34	Polymerizable compounds
35	Oxidizing agents, strong
36	Reducing agents, strong
37	Water and mixtures containing water
38	Water reactive substances

Source: Hatayama, *et al.*, 1980.

chemically similar structure. It must be noted, however, that this is a generally broad method of chemical classification, and that more specific chemical identification should be obtained from such standard reference sources as:

- The Merck Index (Merck 1980),
- Dangerous Properties of Industrial Materials (Sax 1979), or
- Chemical data retrieval services such as CHEMTREC or OHMTADS.

If these sources are inadequate in a specific case, the manufacturer should be contacted.

Once the classes of two chemicals are determined, the matrix in Appendix B may be used to determine the likely reactions resulting from combining the chemicals. It is recognized that numerous variables such as concentration, temperature, and pressure will influence the degree and type of chemical reactions. It is important to note, therefore, that the matrix assumes the chemicals to be of 100% concentration at standard temperature (25°C) and pressure (760 mm Hg). For conditions that vary from these standards, the user is advised to consult the reference sources identified above. The matrix may also be used to determine the compatibility of hazardous wastes if the wastes can be categorized by chemical classes listed in Table 2-1.

The procedure for using the chemical compatibility matrix (Appendix B) is as follows:

1. Determine the chemical classes to which two chemicals belong, as listed in Table 2-1 and Appendix A.

2. Locate the chemical class with the higher number on the left side of the Appendix B chart.

3. Follow that row to the right until it intercepts the column with the lower number.

4. The abbreviation at the point of intersection (explained in the matrix legend) indicates the likely reaction.

5. If the point of intersection is blank, the classes are considered generally compatible. Two or more abbreviations indicate a series of expected reactions in the order in which they would be expected.

As an example, consider determining the compatibility of toluene diisocyanate and nitric acid. From Table 2-1 and Appendix A it is determined that these compounds are in Class 18 (Isocyanates) and Class 2 (Oxidizing Mineral Acids), respectively. Since 18 is the higher number, locate Class 18 on the left side of the matrix and follow that row to the right until it intersects the column for Class 2. The abbreviations "H," "F," and "GT" appear at the point of intersection. Consulting the legend, it is determined that the primary consequences of mixing these two classes of chemicals would be heat generation (H). Secondary consequences resulting from the generation of heat would be fire (F) and generation of toxic gases (GT).

2.3 CHEMICAL/MATERIAL COMPATIBILITY MATRIX

The appropriateness of construction materials for storage tanks and appurtenances is determined on the basis of a variety of objective and subjective factors. Of primary importance is the degree to which the material is resistant to the chemicals to be stored in the tank.

Appendix C is a matrix of compatibility between specific chemicals and a variety of the most commonly used storage tank and appurtenance construction materials. Because corrosion rates vary significantly among chemicals within the same chemical class, the matrix addresses compatibility for specific chemical and material combinations. Table 2-2 is a summary listing of representative chemicals from Appendix A which are specifically listed in the Chemical/Material Compatibility Matrix (Appendix C). These chemicals were chosen because they are commonly encountered in the chemical industry and are functionally representative of the respective chemical classes to which they belong. The list includes most of the top 50 chemicals produced in the United States.

Because corrosion rates are dependent on such factors as concentration, temperature, and humidity, the matrix attempts to identify only the general suitability of a chemical/material combination over a broad range of conditions. Therefore, use of the Appendix C matrix should be limited to a preliminary screening for selection of appropriate materials for a given chemical application. The matrix will also be of value to inspection personnel as a tool for identifying chemical/material gross incompatibilities in existing facilities. For information regarding applicability and corrosion rates under specific conditions, the user is advised to obtain further guidance from qualified design and corrosion engineers, chemical manufacturers, tank fabricators, and standard reference sources (Mellan 1976; Rabold 1951; Staniar 1959; and Cotz 1973).

The materials listed along the horizontal axis of the matrix are those most commonly used in the construction of tanks, valves, appurtenances, and liners. The steels, irons, and aluminums listed in the first seven places are the materials most often used for tank construction. Nickel, monel, inconel, and hastalloys are frequently used in valve and appurtenance applications. Wood and concrete may be used for storage containers or for other applications such as secondary containment or support structures. The remaining materials listed are generally used as tank or valve liners.

To use the Appendix C chemical/material compatibility matrix, find the chemical of interest in the vertical axis. Then follow the row to the right until it intersects the column for the material of interest. The symbol at the point of intersection should be interpreted as follows:

+ = The chemical/material combination is generally suitable under most conditions.

Table 2-2

LIST OF CHEMICAL REPRESENTATIVES BY CLASS

Class 1 Acids, Mineral, Non-Oxidizing

Boric Acid
Chlorosulfonic Acid
Hydriodic Acid
Hydrobromic Acid
Hydrochloric Acid
Hydrocyanic Acid
Hydrofluoric Acid
Hydroidic Acid
Phosphoric Acid

Class 2 Acids, Mineral Oxidizing

Chloric Acid
Chromic Acid
Nitric Acid
Oleum
Perchloric Acid
Sulfuric Acid
Sulfur Trioxide

Class 3 Acids, Organic (All Isomers)

Acetic Acid
Benzoic Acid
Formic Acid
Lactic Acid
Maleic Acid
Oleic acid
Salycilic Acid
Phthalic Acid

Class 4 Alcohols and Glycols (All Isomers)

Allyl Alcohol
Chloroethanol
Cyclohexanol
Ethanol
Ethylene Chlorohydrin
Ethylene Glycol
Ethylene Glyocol Monomethyl Ether
Glycerin
Methanol
Monoethanol Amine

Class 5 Aldehydes (All Isomers)

Acetaldehyde
Formaldehyde
Furfural

Class 6 Amides (All Isomers)

Acetamide
Diethylamide
Dimethylformamide

Class 7 Amines, Aliphatic and Aromatic (All Isomers)

Aminoethanol
Aniline
Diethylamine
Diamine
Ethylenendiamine
Methylamine
Monoethylanolamine
Pyridine

Class 8 Azo Compounds, Diazo Compounds, and Hydrazines

Dimethyl Hydrazine
Hydrazine

Class 10 Caustics

Ammonia
Ammonium Hydroxide
Calcium Hydroxide
Sodium Carbonate
Sodium Hydroxide
Sodium Hypochlorite

Class 11 Cyanides

Hydrocyanic Acid
Potassium Cyanide
Sodium Cyanide

Group 13 Esters (All Isomers)

Butyl Acetate
Ethyl Acetate
Methyl Acrylate
Methyl Formate
Dimethyl Phthalate
Propiolaetone

Class 14 Ethers (All Isomers)

Dichloroethyl Ether
Dioxane
Ethylene Glycol Monomethyl Ether
Furan
Tetrahydrofuran

Class 15 Fluorides, Inorganic

Aluminum Fluoride
Ammonium Fluoride
Fluorosilicic Acid
Fluosilic Acid
Hydrofluorosilicic Acid

Source: Hatayama, *et al.*, 1980.

Table 2-2 (Cont.)

Class 16 Hydrocarbons, Aromatic (All Isomers)

Benzene
Cumene
Ethyl Benzene
Naphthalene
Styrene
Toluene
Xylene

Class 17 Halogenated Organics (All Isomers)

Aldrin
Benzyl Chloride
Carbon Tetrachloride
Chloroacetone
Chlorobenzene
Chlorocresol
Chloroethanol
Chloroform
Dichloroacetone
Dichloroethylether
Dichloromethane (Methylene Dichloride)
Epichlorohydrin
Ethylene Chlorohydrin
Ethylene Dichloride
Freons
Methylchloride
Pentachlorophenol
Tetrachloroethane
Trichloroethylene

Class 18 Isocyanates (All Isomers)

Class 19 Ketones (All Isomers)

Acetone
Acetophenone
Cyclohexanone
Dichloroacetone
Dimethyl ketone
Methyl Ethyl Ketone
Methyl Isobutyl Ketone
Quinone (Benzoquinone)

Class 20 Mercaptans and Other Organic Sulfides (All Isomers)

Carbon Disulfide
Ethyl Mercaptan

Class 21 Metal Compounds, Inorganic

Aluminum Sulfate
Chromic Acid
Silver Nitrate
Tetraethyl Lead
Zinc Chloride

Class 23 Nitriles (All Isomers)

Acrylonitrile

Class 24 Nitro Compounds (All Isomers)

Nitrobenzene
Nitrophenol
Nitropropane
Nitrotoluene
Picric Acid

Class 25 Hydrocarbons, Aliphatic, Unsaturated (All Isomers)

Butadiene
Styrene

Class 26 Hydrocarbons, Aliphatic, Saturated

Butane
Cyclohexane

Class 27 Peroxides and Hydroperoxides Organic (All Isomers)

Benzoyl Peroxide
Hydrogen Peroxide
Chlorocresol
Coal Tar
Cresol
Creosote

Class 28 Phenols, Cresols

Hydroquinone
Nitrophenol
Phenol
Picric Acid
Resorcinol

Class 29 Organophosphates, Phosphothioates, and Phosphodithioates

Malathion
Parathion

Class 31 Epoxides

Epichlorohydrin

Class 32 Combustible and Flammable Materials, Miscellaneous

Diesel Oil
Gasoline
Kerosene
Naphtha
Turpentine

Class 33 Explosives

Benzoyl Peroxide
Picric Acid

Class 34 Polymerizable Compounds

Acrylonitrile
Butadiene
Methyl Acrylate
Styrene

Table 2-2 (Cont.)

Class 35 <u>Oxidizing Agents, Strong</u>

Chloric Acid
Chromic Acid
Silver Nitrate
Sodium Hypochlorite
Sulfur Trioxide

Class 36 <u>Reducing Agents, Strong</u>

Diamine
Hydrazine

Class 37 <u>Water and Mixtures Containing Water</u>

Aqueous solutions and mixtures
Water

Class 38 <u>Water-Reactive Substances</u>

Acetic Anhydride
Hydrobromic Acid
Sulfuric Acid
Sulfur Trioxide

---

c = The chemical/material combination is conditionally suitable, depending upon such factors as temperature, concentration, presence of trace contaminants, degree of agitation, method of material fabrication, etc. More specific data should be obtained from the reference sources cited, to determine suitability under specific conditions.

-- = The chemical/material combination is generally unsuitable under most conditions.

N = Data are insufficient to determine suitability in general. Refer to appropriate references for more specific data.

Selection of the optimum material for a given application is frequently based on economic considerations. However, factors such as the following should also be considered:

- Is the rate of corrosion, even between compatible elements, slow enough that the desired service life is attainable?

- For valves, is the material (particularly liner material) able to withstand pressure changes that might occur during operation?

- For tanks and valves, does the material possess adequate strength characteristics for the design?

- Is an acceptable bond achievable between tank material and liner?

- If a liner material is compatible with the chemical to be stored, what degree of permeability would preclude its use?

In view of such considerations, and understanding that the Appendix C matrix is a summary listing of general compatibility of various chemicals and materials, it can be used as a preliminary materials selection guide. Final selection of materials for specific applications should be made after consulting appropriate professional and technical references.

BIBLIOGRAPHY

Bretherick, L., 1980, Handbook of Reactive Chemical Hazards, CRC Press, Boca Raton, FL.

British Value Manufacturers Association, 1966, Values for the Control of Fluids, Pergamon Press, Elmsford, NY.

Cotz, V., 1972, Plant Engineers Manual and Guide, Prentice-Hall, Englewood Cliffs, NJ.

Evaus, V., 1961, The Corrosion and Oxidation of Metals, St. Martins Press, New York, NY.

__________, 1968, The Corrosion and Oxidation of Metals - First Supplementary Volume, St. Martins Press, New York, NY.

Hatayama, H.K., J.J. Chen, E.R. deVera, R.D. Stephens, and D.L. Storm, 1980, A Method for Determining the Compatibility of Hazardous Wastes, United States Environmental Protection Agency, EPA-600/2-80-076, Washington, D.C.

Hepner, I., ed., 1962, Materials of Construction for Chemical Plant, Leonard Hill Ltd., London.

Hutchinson, J., 1976, ISA Handbook of Control Values, Second Edition, Instrument Society of America, Triangle Park, NC.

Marshall, W., 1981, Construction Materials for Chemical Process Industries, Chemical Engineering, No. 368, New York, NY.

Meidl, H., 1976, Flammable Hazardous Materials, Glencoe Press, New York, NY.

Mellan, I., 1976, Corrosion Resistant Materials Handbook, Noyes Data Corporation, Park Ridge, NJ.

Modern Plastic Handbook, 1981, McGraw-Hill Publications, New York, NY.

National Association of Corrosive Engineers (NACE), 1954, Corrosion Data Survey for Metals, Houston, TX.

__________, 1976, Corrosion Data Survey for Non-Metals, Houston, TX.

__________, 1980, Corrosion Engineer's Reference Book, R.S. Treseder, ed., Houston, TX.

National Fire Protection Association (NFPA), 1981, Hazardous Chemical Reactions, National Fire Code, Section 419M, Volume 14, Boston, MA.

________, 1981, Code for Explosive Materials, National Fire Code, Section 495, Volume 4, Boston, MA.

________, 1981, Identification of Fire Hazards of Materials, National Fire Code, Section 704, Volume 11, Boston, MA.

________, 1981, Classification of Flammable and Combustible Liquids, National Fire Code, Section 321, Volume 3, Boston, MA.

________, 1981, Code for Flammable and Combustible Liquids, National Fire Code, Section 30, Volume 3, Boston, MA.

________, 1981, Fire Hazard Properties of Flammable Liquids, Gases, and Volatile Solids, National Fire Code, Section 325M, Volume 13, Boston, MA.

Pechner, D. and I. Bernstein, 1977, Handbook of Stainless Steel, McGraw-Hill Publications, New York, NY.

Perry, R., and C. Chilton, 1973, Chemical Engineer's Handbook, McGraw-Hill Publications, New York, NY.

Rabald, E., 1968, Corrosion Guide, Elsevier Scientific Publishing Company, New York, NY.

Shreir, L., 1976, Corrosion Volumes One and Two, Newnes-Butterworth Publishing Company, Woburn, MA.

Staniar, W., 1959, Plant Engineering Handbook, McGraw-Hill Publications, New York, NY.

United States Department of Transportation, 1978, CHRIS Hazardous Chemical Data, Washington, D.C.

________, 1980, CHRIS Hazardous Chemical Data - Changes Two, Washington, D.C.

________, 1981, CHRIS Hazardous Chemical Data - Changes Three, Washington, D.C.

3. Storage System Design Elements

3.1 TYPES OF STORAGE TANKS

A general means of classifying storage tanks is by the internal vapor pressure they are designed to sustain. This method of classification transcends the more specific issues of material type and individual applications, such as underground installation, which are discussed in more detail in Sections 3.2 and 3.3, respectively.

Vapor pressure is the pressure exerted on the walls of a closed tank by the vapor contained in the head space above the stored liquid. Due to differences in volatility, vapor pressures vary with temperature. As the temperature of a liquid in a closed tank increases, so will vapor pressure. Volatile chemicals that exhibit relatively high vapor pressures require tanks designed to accommodate those pressures in addition to the static pressures induced by the tank contents.

The vapor pressure criterion yields three categories of tanks: atmospheric tanks, low-pressure tanks, and high-pressure tanks. These three categories are discussed below.

3.1.1 Atmospheric Tanks

Atmospheric tanks are routinely encountered in various industrial settings. The operating pressure is approximately that of atmospheric pressure. Atmospheric tanks are protected by pressure vacuum vents which maintain the pressure difference between the vapor space and the ambient atmosphere at less than a few ounces per square inch. Some typical atmospheric tanks are shown in Figure 3-1. Types of atmospheric tanks include the following:

Coned Roof Tanks may be field-erected to dimensions up to 300 feet in diameter and 64 feet in height. Internal structural support members are frequently employed.

Umbrella Roof and Dome Roof Tanks are variations on the coned roof which employ spherically curved, self-supporting roof plates. They are usually no larger than 60 feet in diameter.

Floating Roof Tanks use a vapor proof seal, floated on the stored liquid and held snug against the tank wall by springs on weights. They are used to eliminate or constantly maintain the vapor space above a stored liquid, and thus minimize filling and vapor expansion losses to the atmosphere.

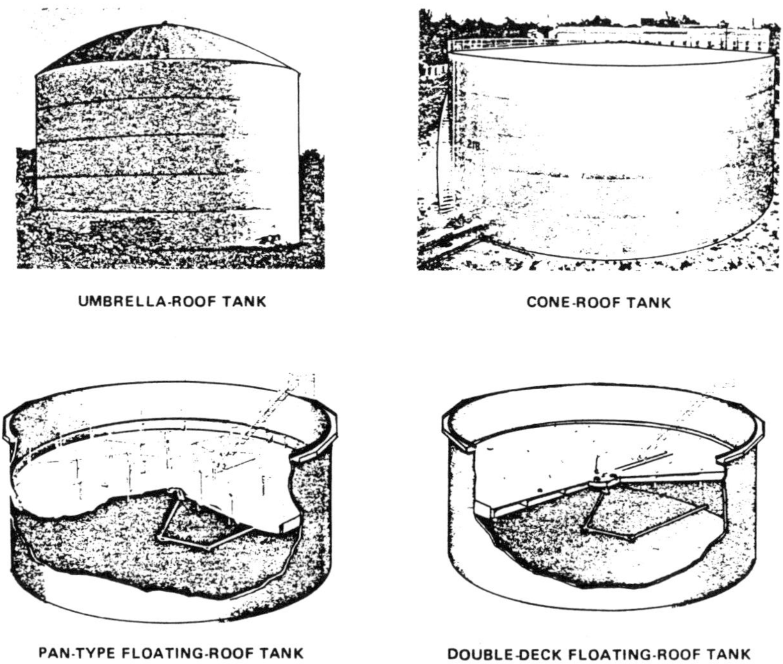

SOURCE: API, 1981

Figure 3–1 TYPICAL ATMOSPHERIC STORAGE TANKS

Fixed Roof - Interior Floater are fixed roof tanks equipped with an interior pan floating roof. They are used when it is desirable to minimize filling and vapor expansion losses, but in climates where snow or rain loads would damage ordinary unprotected floating roofs.

Breather-Roof Tanks employ a flat-roof consisting of a flexible steel membrane capable of expanding and contracting within narrow limits.

Balloon-Roof Tanks are a variation on breather roof tanks, but are capable of greater volume changes.

Vapor-Dome Roof Tanks have a fixed dome in which a flexible interior membrane is free to expand and contract to accommodate large volume changes.

Cylindrical Tanks are used in the vertical or horizontal position to store small volumes of liquids.

3.1.2 Low-Pressure Tanks

Low-pressure tanks are used where the operating pressure ranges from 0.5 pounds per square inch gage (psig) to 15 psig. Pressure relief devices, such as valves or rupture disks, are used to prevent a build-up of pressure beyond the specified safe limit. Some typical low-pressure tanks are shown in Figure 3-2. Types of low-pressure tanks include:

Hemispheroidal, Spheroidal, and Noded Spheroidal Tanks, used for volatile chemicals above 5 psig but less than 15 psig. They are equipped with relief valves to prevent the internal pressure from exceeding the design maximum.

Cylindrical Shell Tanks with coned, domed, or hemispheroidal roofs are used when internal pressures are less than 5 psig but greater than 0.5 psig.

3.1.3 High-Pressure Tanks

High-pressure tanks are used where the operating pressure exceeds 15 psig. Some typical high-pressure tanks are shown in Figure 3-3. Types of high-pressure tanks include:

Pressure Bullet Tanks, horizontal cylindrical steel tanks designed to withstand pressures up to 250 psig. Pressure bullets are primarily used for the storage of compressed gases. Capacity is generally limited to less than 2,000 barrels (bbl).

Spherical Pressure Vessels are used for ambient temperature storage of high vapor pressure materials. In such applications the tank requires shop construction. Field construction is possible for applications where the product vapor pressure is in the 75-psig range for a capacity of up to 50,000 bbl.

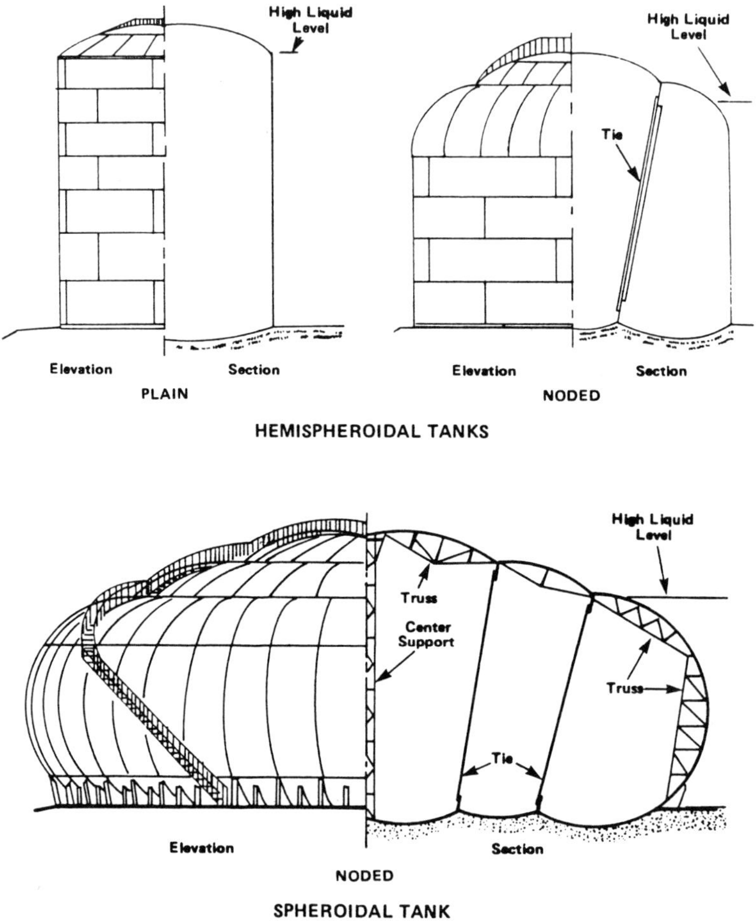

SOURCE: API, 1981.

Figure 3–2 TYPICAL LOW PRESSURE STORAGE TANKS

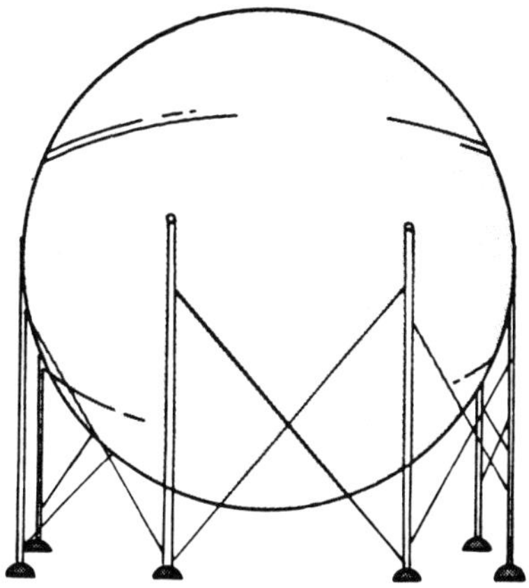

SPHERICAL PRESSURE VESSEL

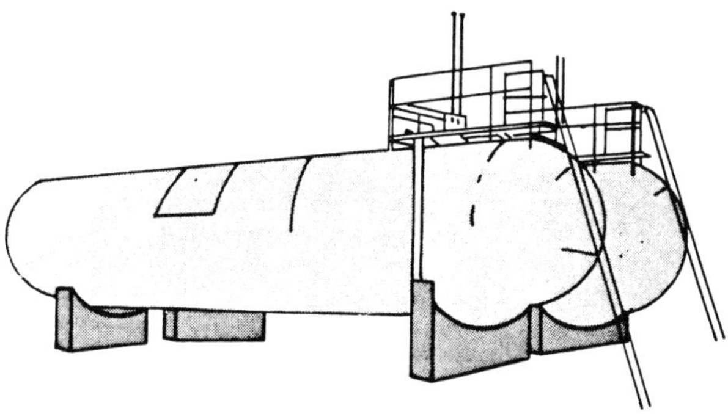

PRESSURE BULLET STORAGE TANK

SOURCE: Higdon, 1976.

Figure 3–3 TYPICAL HIGH-PRESSURE STORAGE TANKS

3.1.4 Selection of Tank Type

By knowing the vapor pressure of a chemical at a given temperature it is possible to specify an appropriate tank type. For example, at 25°C (77°F), pentane exhibits a vapor pressure of nearly 10 psig. A low-pressure tank, which can accommodate pressures ranging from 0.5 psig to 15 psig, is thus suitable in this case. Butane, roughly 35 psig at this temperature, requires a high-pressure tank.

Low-pressure tanks are desirable for volatile chemicals with vapor pressures of less than 15 psig. By keeping vapors confined in a fixed volume, low-pressure tanks keep more of the chemical in the liquid state. By using vapor control devices (Section 3.9) on low-pressure tanks, vapors of a potentially hazardous nature are prevented from venting to the atmosphere. In practice, atmospheric tanks may be utilized for storage of low-volatility chemicals. However, for additional control over vapors, the use of low-pressure tanks is recommended. As previously discussed, high-pressure tanks are required for volatile chemicals with vapor pressures greater than 15 psig.

Table 3-1 gives the recommended tank type for specific chemicals. Selection is based solely on vapor pressure at a single standard temperature, 25°C (77°F). In actual conditions, temperatures may vary considerably, however. Furthermore, the table does not consider the use of vapor control techniques to limit actual operating pressures, the excessive pressures occurring during filling and emptying operations, the implications of cryogenic or heated storage, or other process-specific applications. The table should therefore be used only as a guideline for "red flagging" volatile chemicals which warrant pressure storage considerations. Complete vapor pressure data for listed chemicals can be found in such sources as the Chemical Engineer's Handbook (Perry and Chilton 1973).

Selection of the most appropriate storage system for a given application is dependent upon a multiplicity of factors. The tank type must be able to withstand operating pressures that will occur. Equally important is the selection of construction materials compatible with the stored materials. Guidelines for material selection may be found in Sections 2.3 and 3.2 an Appendix C of this manual, as well as in the references cited in those places. Also important are technical considerations such as tank wall strength and adequacy of foundations and supports. These matters are discussed in Section 3.4. Guidelines for these technical considerations, including valves, ventilation, spill containment, ignition control, and fail-safe devices, are given in the remainder of this chapter. However, ultimate selection of a storage system is dependent upon a number of facility-specific factors, including space available, environmental conditions, economic considerations, desired operating life, etc. For these reasons, determination of the most appropriate storage system should be the responsibility of an engineer familiar with all aspects of the proposed facility.

Table 3-1

STORAGE TANK TYPE FOR
LIQUID CHEMICALS, 25°C (77°F)

Chemical	Tank Type	Chemical	Tank Type
Acetaldehyde	H	Ethylene diamine	A
Acetamide	A	Ethylene dichloride	L
Acetic acid	A	Ethylene glycol	A
Acetone	L	Ethylene glycol monoethyl ether	A
Acetonitrile	L	Formic acid	L
Acetophenone	A	Freons	H
Acrolein	L	Furfural	A
Acrylonitrile	L	Gasoline	A
Allyl alcohol	L	Glycerine	A
Ammonia	H	Hydrocyanic acid	L
Benzene	L	Ispoprene	L
Benzoic acid	A	Methyl acrylate	L
Butane	H	Methyl amine	H
Carbon disulfide	L	Methylchloride	H
Carbon tetrachloride	L	Methyl ethyl ketone	L
Chlorobenzene	L	Methyl formate	L
Chloroethanol	A	Naphtha	A
Chloroform	L	Nitrobenzene	A
Chloropicrin	L	Nitrophenol	A
Chlorosulfonic acid	A	Nitrotoluene	A
Cumeme	A	Pentane	L
Cyclohexane	L	Petroleum oil	A
Cyclohexanane	A	Propane	H
Dichloromethane	L	Pyridine	A
Diesel oil	A	Styrene	A
Diethyl ether	L	Sulfuric acid	A
Dimethylformamide	A	Sulfur trioxide	L
Dimethyl pthalate	A	Tetrachloroethane	A
Dioxane	L	Tetrathydrofuran	L
Epichlorohydrin	A	Toluene	A
Ethanol	L	Trichloroethylene	L
Ethyl acetate	L	Xylene	A
Ethyl benzene	A		

Key: A = Atmospheric, less than 0.5 psig
L = Low Pressure, less than 15 psig but greater than 0.5 psig
H = High Pressure, greater than 15 psig

Source: Ecology and Environment, 1982.

3.2 TANK MATERIALS

As discussed in Section 2.3, it is of primary importance that the material from which a tank is constructed is compatible with the chemicals to be stored within it. Appendix C may be consulted as a guide for determining the general suitability of a variety of chemical and material combinations. Once suitable materials for chemical containment are determined, the inherent strength and economic cost of the materials ultimately determine the material to be used. The general service characteristics and applications of the more common tank construction materials are detailed below.

Mild (Carbon) Steel Tanks. Mild steel is widely used as a tank material due to its strength, durability, ease of fabrication, and relatively low cost. With proper corrosion control techniques, the integrity of a mild steel tank can be enhanced to make it resistant to severe internal and external conditions. For applications which would otherwise be considered incompatible, such as use with dilute sulfuric acid, the mild steel tank can be lined with a variety of resistant materials. Mild steel is appropriate for underground as well as aboveground service. The Steel Tank Institute has developed corrosion control devices and strategies that mitigate previous severe drawbacks of steel tanks.

Reinforced Plastic (RP). Because of their superior chemical resistance, tanks constructed of reinforced plastics, such as fiberglass, have increasingly replaced metallic tanks in particularly severe service applications. Due to their lower strength, however, they are generally limited to use in situations where the operating pressure is close to atmospheric. They should not be used for cryogenic or high temperature applications.

Cylindrical shell RP tanks are predominantly shop fabricated to specification and limited by transportation considerations to 12 feet in diameter and 24 feet in height. Field construction, allowing greater size, is possible. Rectangular RP tanks are extremely susceptible to failure from excessive wall stress, wall deflection, or corner stress. They are difficult to design properly, and usually require considerable horizontal and vertical stiffening.

Stainless Steel. Stainless steel offers superior resistance to chlorinated organics and some acids, as compared to mild steel. Stainless steel is also suitable for very high and low temperature applications. While the cost of stainless steel as a material is high, this may be offset somewhat by the material's high strength.

Aluminum and Aluminum Alloys. Welded aluminum tanks are suitable for use with concentrated nitric acid and sulfuric acid, as well as most organic acids. They should be avoided for use with strong caustic solutions. Aluminum is well suited to low temperature applications. Aluminum alloys also offer good mechanical properties, but a generally lower corrosion resistance.

Other Alloys. Many of the more exotic metal alloys (e.g., monel, hastalloy, inconel, etc.) offer excellent corrosion resistance and strength, but at considerably higher cost. For this reason they are frequently employed in valve applications to meet the demands imposed by excessive pressure and wear.

Concrete. Reinforced concrete is occasionally encountered in atmospheric applications. Its primary drawbacks are high cost, poor chemical resistance, and susceptibility to seepage problems. These can be partially countered by the use of impermeable, resistant liners. An additional drawback is the potential for corrosion of the internal support material, which is usually steel.

3.3 SPECIFIC TANK APPLICATIONS

Atmospheric, low-pressure, and high-pressure tanks can be utilized in a variety of ways to meet specific service requirements. Several of these are detailed below.

3.3.1 Underground Tanks

Underground tanks are generally used for the storage of small or intermediate volumes (1,000 to 20,000 gallons) of gasoline, fuel oils, or a variety of chemicals. They should be constructed from steel with fiberglass or cathodic protection, or from reinforced plastic. Their characteristics and weaknesses are further discussed in Section 5.2.2.

Underground steel tanks require special installation considerations to insure adequate support and stability, as well as to minimize corrosion. Although installation criteria should be determined on a facility-specific basis, the following criteria are offered as guidelines. The excavation for underground tanks should extend at least one foot in all directions from the in-place tank profile. The base of the excavation should be laid with a backfill bed of at least 12 inches of non-corrosive material such as pea gravel, sand, or No. 8 crushed stone. To minimize corrosion, the surrounding soil should have a resistivity of at least 10,000 OHM-per-cm. In soils with a lower resistivity, cathodic protection should be employed, or reinforced plastic tanks should be selected. (Note: Many states require cathodic protection of underground steel tanks, even in situations where soil resistivity may exceed 10,000 OHM-per-cm. The user of this manual is advised to consult and comply with all rules and regulations governing underground tank installation applicable within a specific state or local jurisdiction.) The backfill should extend to at least 12 inches above the top of the buried tank. At a minimum, the tank must be covered by at least two feet of earth, or one foot of earth and four inches of concrete or asphalt pad extending one foot beyond the tank perimeter. If the surface will be subjected to traffic, this depth should be increased to at least three feet of earth, or 18 inches of earth and eight inches of concrete. In all cases, installation should be determined on an individual basis, and should be consistent with any applicable manufacturer's directions. Additional guidelines are contained in NFPA 30, Flammable and Combustible Liquids Code.

Horizontal tanks designed for atmospheric service can be used in underground applications and have several desirable features. The safety of underground tanks against accidental damage is greatly enhanced, while the need for diking is eliminated. This advantage, however, may be offset in severe applications, where underground containment in the form of clay, concrete, or plastic liners may be needed. Tank burial has a temperature moderating effect that will prevent solidification of contents in some cases, and reduce vapor generation. Structural support problems encountered with horizontal aboveground tanks are eliminated by the uniform support achieved with underground installations. A savings in ground space or building space is also an advantage.

A major drawback to underground tanks is that, due to their inaccessibility, structural or corrosion problems are not readily apparent. This is discussed further in Section 5.2.2.

3.3.2 Cryogenic Tanks

Cryogenic tanks are generally used to maintain hydrocarbon gases, such as liquefied natural gas (LNG) or liquid propane (LPG). They are designed according to Section VIII of the ASME Boiler and Pressure Vessel Code.

Most cryogenic tanks have vacuum-jacketed insulation, although insulation can be accomplished in other ways. An insulating blanket may be applied to the exterior of a single-wall tank, or a layer of insulating material such as perlite may be contained between the walls of a double-wall tank. The inside tank must meet the temperature and pressure requirements, while the outside wall acts only as a containment wall. Thus, the inside tank is generally constructed of nickel or aluminum alloy, copper, or 304 stainless steel, and the outside tank is constructed of ordinary carbon steel. Adequate maintenance of the tank insulation is essential, and all joints should be vacuum tight. Piping between the two walls of a vacuum-insulated tank should be long and flexible, and is usually copper tubing. Valving should be of the extended stem type. Further, electrically heated foundations are frequently employed to reduce the impact of frost heaves on the tank.

Appendix Q of API Standard 620, which is specifically applicable to liquefied hydrocarbon gases, can provide guidelines for low-temperature service to -270°F.

3.3.3 Heated Tanks

Heated tanks are frequently used to maintain the desired viscosity of petroleum-derived asphalt products and crude oils, to prevent phase change of the product, and for other applications. Heat is applied through the use of steam lines, plate coils, or heat exchangers, etc. Mixing of tank contents may be necessary to maintain uniform temperatures, and may be achieved by installation of mechanical mixers and pumps.

Temperatures in a heated storage tank should be monitored to prevent the contents from undergoing a thermal phase change. If the temperature is not monitored and it rises to a point above the recommended temperature range for the product, evaporation could occur and the resulting pressure increase could lead to an explosion. Similarly, if the temperature drops and solidification occurs, product expansion could result in the storage tank bursting.

Where insulation is applied to the exterior of a tank, it must be adequately protected from the elements and properly maintained. Adequate pressure-relieving capacity is also required. Conventional design criteria apply up to temperatures of 200°F, with additional considerations in effect at higher temperatures (see Appendix M of API standard 650). Open-top tanks and floating-roof tanks are not suitable for heated service.

3.4 TANK DESIGN CRITERIA

Proper design of a storage system involves the consideration of numerous physical and chemical criteria. These include:

- Compatibility of the tank material with the material to be stored;
- Specific gravity of the liquid to be stored;
- Desired volumetric capacity of the tank;
- External loads on the tank, such as wind loads;
- Use of required and optional appurtenances;
- Proper preparation of the tank installation area;
- Static pressure induced by the tank contents; and
- Service pressure of the tank.

Each of these criteria should be defined prior to tank design and given appropriate attention during the design process.

A liquid's specific gravity is a major consideration during the structural design process because it is the determining factor in calculating the tank shell and foundation stresses. For these reasons, a tank designed for a liquid of a given specific gravity should not be used indiscriminately for liquids of greater specific gravity. In cases where it is necessary to store a heavier liquid than the tank was designed for, calculations should be performed to determine a fill height that would prevent stresses in excess of those for which the tank was originally intended. As discussed in Section 3.1, the structural design must also account for the vapor pressures induced by the tank contents.

It is important to note that underground tanks should be designed to be able to bear all external loads to which they may be subjected.

Due consideration should be given to traffic loads if the surface under which the tank is installed is subjected to traffic.

Various guidelines for the design of different types of storage tanks have been promulgated by independent trade organizations and professional societies. While such guidelines are not legally or technically binding, they provide the reasonable standards and specifications for proper design. They should not take the place of sound engineering reasoning in specific situations that require additional design considerations. A listing of these guidelines is contained in Table 3-2.

3.4.1 Wall Thickness Design

Although the desired volumetric or length capacity of a storage tank is a function of its diameter and height (vertical tanks) or length (horizontal tanks), the required wall thickness of a liquid storage tank is calculated from the liquid's specific gravity, the desired tank capacity and dimensions, and internal head space pressure. In general, the heavier a liquid or larger the diameter of the tank, the greater the wall thickness that will be required for structural stability. Table 3-3 illustrates volume changes brought about by changes in tank diameter.

Minimum wall thickness for steel or other metallic vertical tanks is generally set at 3/16 inch, although 1/4 inch is usually more desirable. Minimum thicknesses for atmospheric and low-pressure steel tanks, which are vertical and cylindrical, are contained in Table 3-4. These values are exclusive of any corrosion allowance specified in the appropriate design standards.

For horizontal steel tanks, recommended wall thicknesses are listed in Table 3-5. It is also recommended that the length of a horizontal tank not exceed six times its diameter.

Thickness may also be variable with height along the sides of a vertical tank, with the lower cross sections requiring greater thickness than the upper. This approach to design is referred to as graduated wall thickness, and is frequently employed in shop-fabricated, reinforced plastic tanks. Table 3-6 outlines recommended minimum thicknesses for graduated wall, reinforced plastic tanks. A safety factor of 10 is built into these recommendations. A liquid specific gravity of 1.2 is assumed.

In addition to structural considerations in calculating wall thickness, it is necessary to provide additional thickness in applications where corrosion is suspected or possible. A corrosion allowance should be established based upon the rate of corrosion, if known, and the desired service life of the tank. When the rate of corrosion is unknown or is variable in magnitude and extent, a minimum corrosion allowance thickness should be applied. This is discussed further in Section 5.5.

Table 3-2

EXISTING STRUCTURAL GUIDELINES

Tank Type	Existing Guidelines	Promulgating Organization	Comment
High Pressure	Boiler and Pressure Vessel Code Section VIII, Divisions 1 and 2 Section X, fiberglass reinforced plastic pressure vessels	American Society of Mechanical Engineers 345 E. 47th Street New York, NY 10017 212/705-7722	
Low Pressure	Standard 620, recommended rules for design of large, welded, low-pressure storage tanks	American Petroleum Institute 201 L Street, NW Washington, DC 20057 202/457-7000	Applicable to non-petroleum as well as petroleum storage tanks Sections VIII and X of the ASME Boiler and Pressure Vessel Code also apply
Atmospheric	Standard 650, welded steel for oil storage	API	Applicable to non-petroleum as well as petroleum storage tanks
	Standard 12A, oil storage tanks with riveted shells	API	
	Standard 12B, bolted production tanks	API	
	Standard 12D, large welded production tanks	API	
	Standard 12E, wooden production tanks	API	
	Standard 12F, small welded production tanks	API	

Table 3-2 (Cont.)

Tank Type	Existing Guidelines	Promulgating Organization	Comment
Atmospheric (Cont.)	Standard for welded aluminum-alloy storage tanks, ANSI B96.1 - 1981	American National Standards Institute, Inc. 1430 Broadway New York, NY 10018	
	Standard steel tanks, D100-67	American Water Works Association 6666 W. Quincy Ave. Denver, CO 80234 303/794-7711	Adaptable to storage tanks of chemicals as well as water
	Steel underground tanks for flammable and combustible liquids, UL 58	Underwriters Laboratory 333 Pfingsten Rd. Northbrook, IL 60062 312/272-8800	
	Steel above-ground tanks for flammable and combustible liquids	Same as above	

Source: Ecology and Environment, Inc., 1982.

Table 3-3

GALLON CAPACITY PER FOOT OF HEIGHT OR LENGTH IN CYLINDRICAL TANKS

Diameter (feet)	Gallons per Foot of Height (Length)
5.0	146.88
5.5	177.72
6.0	211.51
6.5	248.23
7.0	287.88
7.5	330.48
8.0	376.01
8.5	424.48
9.0	475.89
9.5	530.24
10.0	587.52
10.5	647.74
11.0	710.90
11.5	776.99
12.0	846.03
12.5	918.00
13.0	992.91
13.5	1070.8
14.0	1151.5
14.5	1235.3
15.0	1321.9
15.5	1411.5
16.0	1504.1
16.5	1599.5
17.0	1697.9
17.5	1799.3
18.0	1903.6
18.5	2010.8
19.0	2120.9
19.5	2234.0
20.0	2350.1
20.5	2469.1
21.0	2591.0
21.5	2715.8
22.0	2843.6
22.5	2974.3
23.0	3108.0
23.5	3244.6
24.0	3384.1
24.5	3526.6
25.0	3672.0
25.5	3820.3
26.0	3971.6
26.5	4125.9
27.0	4283.0
27.5	4443.1
28.0	4606.2
28.5	4772.1
29.0	4941.0
29.5	5112.9
30.0	5287.7
30.5	5465.4

Source: Ecology and Environment, Inc., 1983.

Table 3-4

VERTICAL STEEL TANK
MINIMUM WALL THICKNESSES*

Tank Diameter (feet)	Thickness (inches)
Smaller than 50	3/16
50 to 120, Excl.	1/4
120 to 200, Incl.	5/16
Over 200	3/8

*Exclusive of any corrosion allowance or variations in liquid density of tank contents.

Source: API, 1978.

Table 3-5
HORIZONTAL STEEL TANK MINIMUM WALL THICKNESSES*
Thickness of Steel Underground Tanks

Capacity		Maximum Diameter		Manufacturers' Standard or Galvanized Sheet	Nominal Thickness			
					Uncoated		Galvanized	
U.S. Gallons	dm^3	Inches	m	Gage No.	Inches	mm	Inches	mm
Up to 285	Up to 1,078	42	1.07	14	0.075	1.91	0.079	2.01
286 to 560	1,082 to 2,120	48	1.22	12	0.105	2.67	0.108	2.74
561 to 1,100	2,124 to 4,164	64	1.63	10	0.135	3.43	0.138	3.51
1,101 to 4,000	4,168 to 15,142	84	2.13	7	0.179	4.55		
4,001 to 12,000	15,145 to 45,425	126	3.20	1/4 inch	0.250	6.35		
12,001 to 20,000	45,429 to 75,708	144	3.66	5/16 inch	0.312	7.92		
20,001 to 50,000	75,712 to 189,270	144	3.66	3/8 inch	0.375	9.53		

Source: Underwriters Laboratory, Inc., 1976.

Thickness of Steel Aboveground Tanks

Capacity U.S. Gallons (kl)		Maximum Diameter, Inches (m)		Minimum Metal Thickness, Inches (mm)			
				Carbon Steel		Stainless Steel	
550 or less	(2.13)	48	(1.22)	0.093	(2.36)	0.071	(1.80)
551 to 1,100	(2.14 to 4.26)	64	(1.63)	0.123	(3.12)	0.086	(2.18)
1,101 to 9,000	(4.27 to 34.87)	76	(1.93)	0.167	(4.24)	0.115	(2.92)
1,101 to 35,000	(4.27 to 135.63)	144	(3.66)	0.240	(6.10)	0.158	(4.01)
35,001 to 50,000	(135.64 to 193.77)	144	(3.66)	0.365	(9.27)	0.240	(6.10)

Source: Underwriters Laboratory, Inc., 1981.

Table 3-6

REINFORCED PLASTIC TANK
MINIMAL GRADUATED WALL THICKNESSES*

Distance From Top of Tank (feet)	Minimum Wall Thickness (inches) for Tanks of Diameter:														
	2 (feet)	2 1/2 (feet)	3 (feet)	3 1/2 (feet)	4 (feet)	4 1/2 (feet)	5 (feet)	5 1/2 (feet)	6 (feet)	7 (feet)	8 (feet)	9 (feet)	10 (feet)	11 (feet)	12 (feet)
2	3/16	3/16	3/16	3/16	3/16	3/16	3/16	3/16	3/16	3/16	3/16	3/16	3/16	3/16	3/16
4	3/16	3/16	3/16	3/16	3/16	3/16	3/16	3/16	3/16	3/16	3/16	3/16	3/16	3/16	3/16
6	3/16	3/16	3/16	3/16	3/16	3/16	3/16	3/16	3/16	3/16	3/16	3/16	1/4	1/4	1/4
8	3/16	3/16	3/16	3/16	3/16	3/16	3/16	3/16	3/16	1/4	1/4	1/4	1/4	1/4	5/16
10	3/16	3/16	3/16	3/16	3/16	3/16	3/16	1/4	1/4	1/4	1/4	1/4	5/16	5/16	5/16
12	3/16	3/16	3/16	3/16	3/16	3/16	1/4	1/4	1/4	1/4	1/4	5/16	5/16	5/16	3/8
14	3/16	3/16	3/16	3/16	1/4	1/4	1/4	1/4	1/4	5/16	5/16	5/16	5/16	3/8	3/8
16	3/16	3/16	3/16	1/4	1/4	1/4	1/4	1/4	1/4	5/16	5/16	3/8	3/8	3/8	7/16
18	3/16	3/16	3/16	1/4	1/4	1/4	1/4	5/16	5/16	5/16	3/8	3/8	3/8	7/16	1/2
20	3/16	3/16	1/4	1/4	1/4	1/4	5/16	5/16	5/16	3/8	3/8	3/8	7/16	1/2	1/2
22	3/16	1/4	1/4	1/4	1/4	5/16	5/16	5/16	5/16	3/8	3/8	7/16	1/2	1/2	9/16
24	3/16	1/4	1/4	1/4	1/4	5/16	5/16	5/16	3/8	3/8	7/16	1/2	1/2	9/16	5/8

*Exclusive of any corrosion allowance or variations in liquid density of tank contents.

Source: Mallinson, 1969.

3.4.2 Foundations

Proper foundation design and construction should accomplish two goals:

1. Provide uniform and adequate support to the full weight of the tank and its contents; and

2. Avoid creation of localized sites susceptible to corrosion.

Large tanks resting directly on the ground should be underlain by a minimum of four inches of oil-treated sand or other pervious, well graded soil to provide flexible, continuous support across the entire bottom plate as well as to promote drainage away from the underside of the bottom plate. The sub-grade should be free of surface pockets of loam or organics-containing topsoil, and should be of adequate strength to support the weight of the tank when full. Approximate tank load can be calculated (neglecting the contribution of the structural material itself) by multiplying the density of the liquid (pounds per cubic foot) by the height of the tank (feet). This value should not exceed the bearing capacity of the local soil. The approximate bearing capacities of common soil types are given in Table 3-7. The actual bearing capacity of the supporting soils should be determined by a proper soils investigation.

A schematic of a good tank foundation for large outside tanks is provided in Figure 3-4. The foundation sealer and adequate drainage grading are important to prevent the accumulation of precipitation around the tank foundation and to minimize moisture under the tank.

For large tanks with high shells, a foundation ringwall may be required. A ringwall serves to better distribute the tank load and creates a more uniform soil-loading condition. This is especially desirable when the bearing capacity of the underlying soil alone is marginally acceptable. A cross-sectional view of a ringwall foundation is contained in Figure 3-5. Complete specifications for such foundations are contained in API 650.

Foundations for underground tanks should be designed to support the tank and its contents plus any superimposed loads. Consideration should be given to uplift forces if the tank is located in an area with a high groundwater table or subject to flooding. If the tank is subject to uplift, holddown straps should be provided to resist the uplift forces.

3.4.3 Supports

Smaller tanks resting on structural supports are subject to the same requirements as earth-supported tanks. For adequate support, horizontal cylindrical tanks should rest on saddles that make contact on at least 120° of their circumference. To minimize potential point sources of corrosion, the ends and edges of these saddles should be angled to allow drainage of precipitation or spillage away from the tank surface. Contact should ideally consist of a metal reinforcing wear plate hermetically sealed to the tank and a metal saddle, which

Table 3-7

APPROXIMATE BEARING CAPACITIES

Soil Type	Tons/Square Foot
Soft clay	1
Dry fine sand	2
Dry fine sand with clay	3
Coarse sand	3
Dry hard clay	3.5
Gravel	4
Rock	10 to 40

Source: Perry and Chilton, 1973.

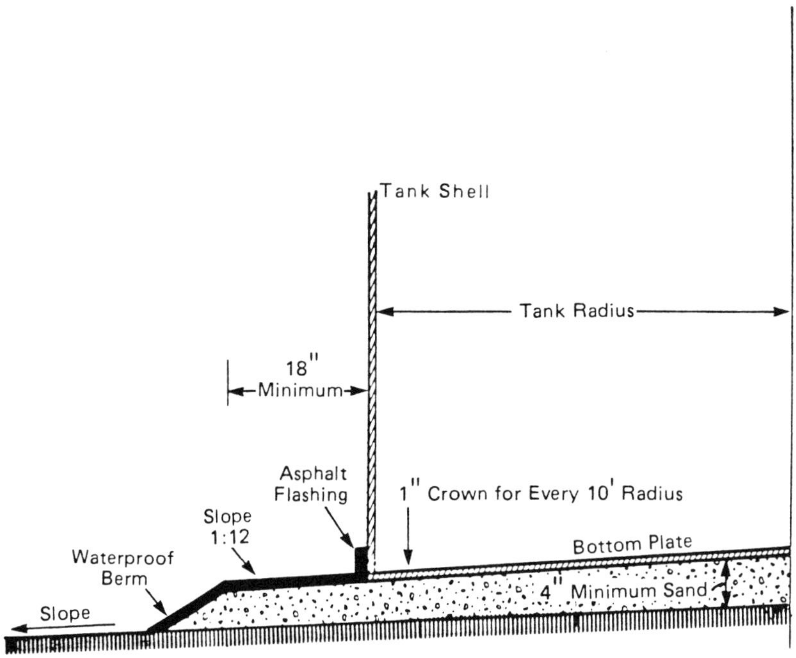

SOURCE: Staniar, 1959.

Figure 3–4 FOUNDATION SCHEMATIC FOR OUTDOOR TANKS

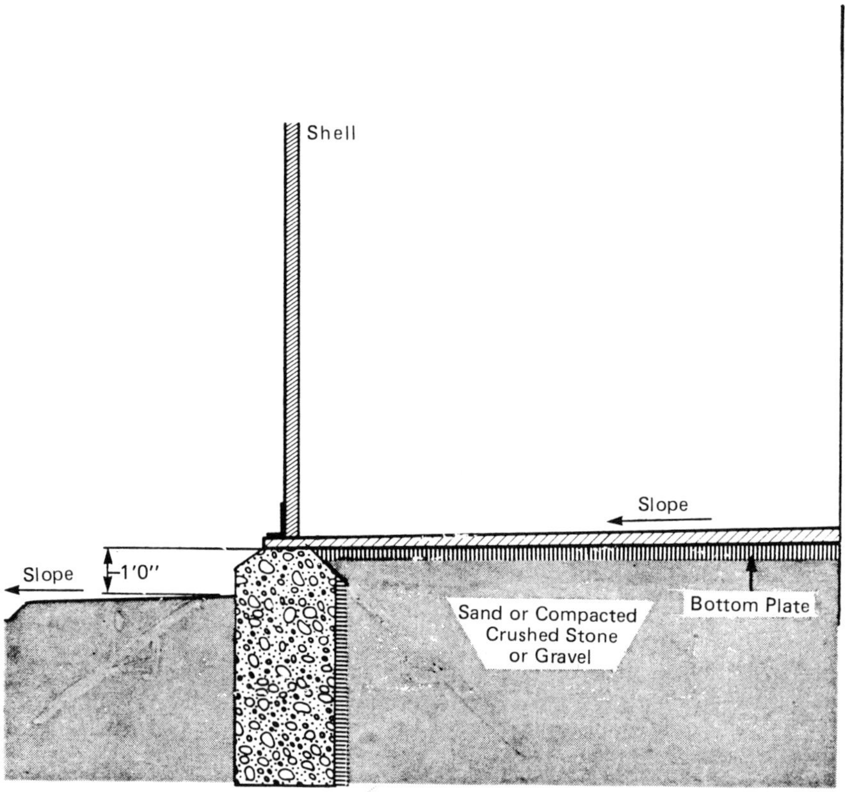

SOURCE: Staniar, 1959.

Figure 3–5 ILLUSTRATIVE RING WALL FOUNDATION

in turn rests on a concrete pier. Alternatively, although less desirably, the plate may be sealed to the tank and rest directly on a concrete saddle. In no instances should the wear plate consist of decomposable material such as tar-saturated felt paper, as this provides a moist surface to encourage corrosion.

Similarly, the supports for mounted vertical tanks should be continuously sealed along all points of contact and tapered away from exposed surfaces. Figure 3-6 illustrates these points. The undersides of vertical tanks should receive good air circulation and be accessible for visual inspection and routine maintenance.

3.5 VALVE SELECTION

Selection of a specific valve to be used in conjunction with a liquid storage tank should be based on that liquid's viscosity and corrosivity, as well as its temperature and velocity. The expected service pressure is used to select the appropriate valve class, as established by ANSI, Standard B34. An additional consideration is whether the valve will be used for merely isolating the flow (open and close), or for the more demanding task of regulating the velocity or rate of flow.

The choice of valve material should be based on the aforementioned temperature and pressure considerations, as well as the corrosive properties of the liquid being handled. The most common metallic materials include cast iron, bronze, nickel alloys, steel, stainless steel, aluminum, and titanium. Cast iron and bronze are generally used for applications up to 260°C, with carbon steel and the alloy steels used at higher temperatures. Nickel steels are used in application down to -57°C. Chemically resistant valve liners include butyl rubber, neoprene, polyethylene, polyvinyl chloride (PVC), and glass. Reference can be made to Appendix C to determine the general compatibility of a specific liquid with a material; however, final selection should be based on the expert advice of a reputable valve manufacturer and the design engineer.

Minimum thickness requirements for valve bodies has been established by the American National Standards Institute (ANSI; Standard B16.5.) The recommended thickness is set at 1.5 times the thickness that would be calculated for a simple cylinder with a maximum allowable stress of 7,000 pounds per square inch (psi), and subjected to an internal pressure equal to the expected maximum on-line service pressure. The safety factor of 1.5 compensates for unquantifiable effects of internal turbulence and stress concentrations. Minimum thicknesses are further discussed in Section 5.5.

The applications and limitations of the major valve types are listed below.

Gate Valves. Gate valves should be used to isolate flow through a line. They are not suitable for use as a flow regulator or throttling device. In cases where they must be so used, they should be replaced with a more suitable alternative such as a globe valve.

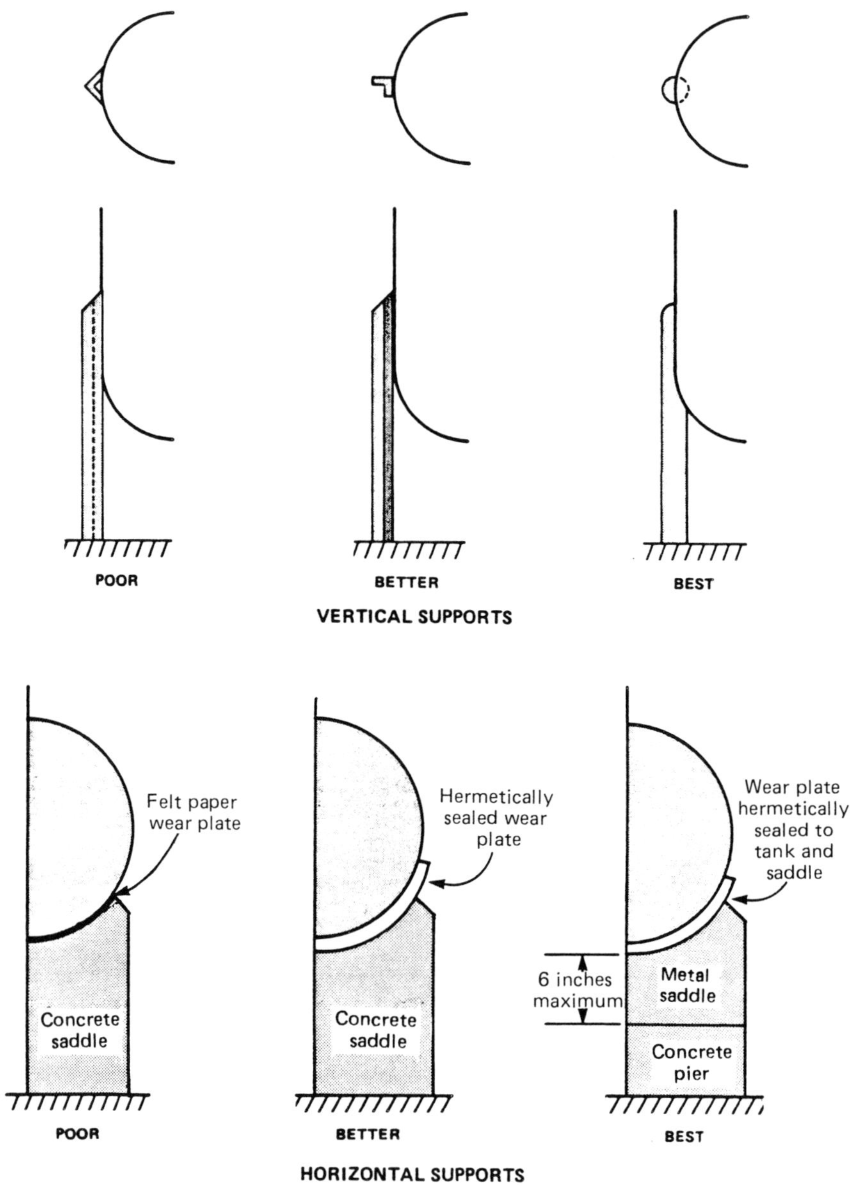

SOURCE: Staniar, 1959

Figure 3–6 ASSESSMENT OF VERTICAL AND HORIZONTAL TANK SUPPORTS

Globe Valves. Globe valves are well suited for throttling or flow-regulating applications. They are usually installed with the stems in a vertical position. This position, however does not allow complete drainage from the valve, which may be of concern in corrosive applications. The pressure drop across the globe valve is greater than other valve types.

Diaphragm Valves. Diaphragm valves are well suited to regulating service, but are limited to pressures of 50 psi. They are also ideal under corrosive or reactive conditions, since various synthetic liners and diaphragms are available. They should not be used in the presence of a vacuum, as this may cause separation of the diaphragm from the housing.

Butterfly Valves. Butterfly valves are also well-suited to regulating service and isolating service. They are usually used in low-pressure service for coarse flow control, but are available for use at higher pressures.

Safety (Pressure Relief) Valves. Safety, or pressure relief, valves should be required of all atmospheric, low-pressure, and high-pressure tanks to insure that the safe pressure for that tank is not exceeded. Such valves are used to quickly vent the vapors in the head space of a tank and should not come in contact with the liquid fraction itself. They should be installed in such a way that they release to a point of safe discharge.

Check (Back Flow) Valves. Check valves are used to prevent the reversal of flow through a pipe, or back into a tank, for example from a discharge line. Swing check valves are used in horizontal lines; lift check valves are used in vertical lines where the flow is upward; globe check valves are used in horizontal lines; angle check valves are used where a vertical line with upward flow turns horizontal; and tilting-disk check valves are used in horizontal lines or vertical lines where the flow is upward. The tilting-disk variety generally provides the fastest closing action. Globe and angle check valves normally incorporate an integral dashpot above the disk to slow the motion of the disk and reduce wear. Check valves in general are more likely to leak at low pressure than other valve types, since it is high pressure that insures their tight closing seal.

Ball Valves. Ball valves are used for regulating service, and are available for high- or low-pressure applications. However, they are limited to use with temperatures and fluids which the valve seats will resist. Ball valve seats are available in a variety of materials, including Teflon, polyethylene, nylon, Monel, hastalloy, and stainless steel.

Critical areas of these valves are treated in Section 5.2.5.

3.6 TANK VENTING AND CONTROL OF VAPOR EMISSIONS

Venting or vapor emission considerations apply in one form or another to all liquid storage tanks. The application of vents and vapor emission control devices depends on tank size and construction, liquid vapor pressure, and filling and emptying rates. In general, tank vents and vapor emission controls are required to compensate for the following conditions:

- Air intake during tank emptying;
- Vapor exhaust which occurs during tank filling;
- Expansion and contraction of the vapor space due to temperature fluctuations (breathing losses);
- Normal evaporation of the liquid; and
- Emergency situations, such as fires.

Inadequate venting systems can result in tank failure due to excessive pressure or vacuum buildup.

3.6.1 Normal Vents

Normal vents apply to atmospheric and low-pressure tanks which are not constructed to handle excessive pressure or vacuum build-up. High-pressure tanks require emergency vents only. Capacity requirements for normal vents are specified in terms of cubic feet of free air per hour (CFH) for each 100 barrels (4,200 gallons) per hour of maximum emptying or filling rate. API Standard 2000 (Venting Atmospheric and Low-Pressure Storage Tanks) specifies venting capacities of 560 CFH for tank emptying, 600 CFH for tank filling, and additional capacities based on the liquid flash point and the tank size for thermal inbreathing and outbreathing. These additional specifications are given in Table 3-8. For example, a 126,000-gallon tank would require 560 CFH vent capacity for emptying; 600 CFH for filling; 3,000 CFH for thermal inbreathing; and either 1,800 CFH or 3,000 CFH for thermal outbreathing, depending on the vapor flash point. The vent capacities are additive; that is, the venting specified for filling and emptying is not sufficient to handle thermal breathing, even if the vent capacity for filling and emptying is higher than that for thermal breathing.

Venting under normal operating conditions can be achieved with open vents, pressure vacuum (PV) valves, pressure relief valves, and pilot-operated relief valves, and each is generally designed for specific services. Open vents without a flame arresting device, such as a metal screen, should only be used for liquids with flash points above 100°F or for tanks with capacities less than 2,500 gallons. Pressure vacuum valves are designed for atmospheric storage tanks containing low-boiling liquids such as petroleum liquids. According to the API _Guide for Inspection of Refinery Equipment_, pressure relief valves are used chiefly in liquid storage, and generally should not be used in conjunction with gas or vapor service. Rupture discs and resilient valve seats are often used in conjunction with pressure relief valves in the storage of corrosive, viscous, and polymerizable liquids which can damage the valve. Pilot-operated valves are generally applied where the relief pressure is near the operating pressure and for low-pressure storage tanks. They should not be used in

Table 3-8

REQUIREMENTS FOR THERMAL VENTING CAPACITY

Tank Capacity	Thermal Venting Capacity (cubic feet of free air per hour)		
	Inbreathing (Vacuum)	Outbreathing (Pressure)	
Gallons		Flash Point > 100°F (37.78°C)	Flash Point < 100°F (37.78°C)
2,500	60	40	60
4,200	100	60	100
21,000	500	300	500
42,000	1,000	600	1,000
84,000	2,000	1,200	2,000
126,000	3,000	1,800	3,000
168,000	4,000	2,400	4,000
210,000	5,000	3,000	5,000
420,000	10,000	6,000	10,000
630,000	15,000	9,000	15,000
840,000	20,000	12,000	20,000
1,050,000	24,000	15,000	24,000
1,260,000	28,000	17,000	28,000
1,470,000	31,000	19,000	31,000
1,680,000	34,000	21,000	34,000
1,890,000	37,000	23,000	37,000
2,100,000	40,000	24,000	40,000
2,520,000	44,000	27,000	44,000
2,940,000	48,000	29,000	48,000
3,360,000	52,000	31,000	52,000
3,780,000	56,000	34,000	56,000
4,200,000	60,000	36,000	60,000
5,040,000	68,000	41,000	68,000
5,880,000	75,000	45,000	75,000
6,720,000	82,000	50,000	82,000
7,560,000	90,000	54,000	90,000

Source: API, 1982.

conjunction with viscous liquids or with liquids whose vapors can polymerize. Vapors emitted during transfer operations may occur as strong emissions from the transfer lines or transport vehicle, or they may arise due to vapor expansion within the tank being filled. Transfer lines may be equipped with vapor recovery systems which reroute vapors back to the transport tank. Expanding vapors within the storage tank may be best controlled if the tank is equipped with a floating roof. Otherwise, the vapors may be routed from the tank to carbon adsorption units, thermal and catalytic incinerators, or refrigerated condensors. Control of vapor emissions is further discussed in Section 3.6.3.

3.6.2 Emergency Vents

Emergency vents are designed to safeguard against the potential rapid evaporation of the stored liquid. The most common cause of such emergencies is exposure of a tank to fire. One common safeguard is to design the tank so that the weld at the roof-to-shell attachment will fail preferentially to welds which could cause the liquid contents of the tank to be discharged if they failed. For tanks without weak roof-to-shell attachments, larger or additional normal vents, and gage hatches or manhole covers which open at designated pressures, can be used for emergency venting. The additional venting capacities required for fire exposure are given as a function of wetted tank surface area in Table 3-9.

Safety-relief valves are designed primarily for use in flammable or toxic materials service. They are often connected to a piping system which reroutes the discharge to an appropriate remote discharge point or to a control device such as a recovery unit, flare, or carbon adsorption bed.

3.6.3 Vapor Emissions Control Schemes

Four basic tank design schemes can be used to control volatile chemical emissions which can contribute to health and environmental problems. These tank categories are fixed roof, floating roof, variable vapor space, and pressure tanks. The selection of the appropriate scheme is basically a function of vapor pressure and tank size. Federal and state agencies have developed regulations based on vapor pressure which designate control categories for tanks greater than 40,000 gallons. In general, vapor emission control is not required on tanks of less than 40,000 gallons capacity.

Fixed roof tanks offer the minimum acceptable control of vapor emissions from volatile chemicals. Pressure vacuum (PV) vents are standard accessories on fixed roof tanks and are designed to allow pressure variance over a range of only -0.03 psig to +0.03 psig. Vapor loss occurs whenever these limits are surpassed. Fixed roof tanks are most applicable for storing slightly to moderately volatile chemicals in quantities less than 40,000 gallons.

Fixed roof tanks can provide complete control of vapor emissions if equipped with a vapor recovery system. Vapor recovery systems function by collecting the vapors, usually in a manifold attached to

Table 3-9

TOTAL RATE OF EMERGENCY VENTING REQUIRED FOR FIRE EXPOSURE VERSUS WETTED SURFACE AREA (NONREFRIGERATED ABOVEGROUND TANKS)

Wetted Area (square feet)	Venting Requirement (cubic feet of free air per hour)	Wetted Area (square feet)	Venting Requirement (cubic feet of free air per hour)
20	21,100	350	288,000
30	31,600	400	312,000
40	42,100	500	354,000
50	52,700	600	392,000
60	63,200	700	428,000
70	73,700	800	462,000
80	84,200	900	493,000
90	94,800	1,000	524,000
100	105,000	1,200	557,000
120	126,000	1,400	587,000
140	147,000	1,600	614,000
160	168,000	1,800	639,000
180	190,000	2,000	662,000
200	211,000	2,400	704,000
250	239,000	2,800	742,000
300	265,000	>2,800	742,000

Source: API, 1982.

more than one tank, and recovering them through vapor/liquid absorption, compression, refrigeration or vapor/solid adsorption. Recovered vapors can also be incinerated, or otherwise destroyed. A schematic diagram of a typical vapor recovery unit is given in Figure 3-7. Table 3-10 lists several examples of chemicals whose vapors are commonly recovered.

Floating roof tanks provide greater vapor emissions control than fixed roof tanks with no vapor recovery. The greatest potential vapor loss from floating roof tanks is from improper fitting of the seal and the shoe to the shell. Additional vapor loss can occur as the roof descends and residual liquid on the tank wall evaporates. In cases where a floating roof is installed inside a fixed roof tank, venting of the space between the floating and fixed roofs is required to prevent the formation of explosive mixtures. Floating roof tanks or the equivalent offer the minimum acceptable control of vapor emissions for liquids having vapor pressures between 1.5 pounds per square inch absolute (psia) and 11.0 psia and stored in quantities greater than 40,000 gallons.

Variable vapor space tanks provide an expanding and contracting vapor space, thus allowing pressure changes in the tank without the need for narrow-range PV vents. They are commonly used in conjunction with more than one tank. The two most common variable vapor space systems are those employing flexible diaphragms and those with lifter roofs. Flexible diaphragms are installed in gasholder units which are either mounted directly on a tank (see Figure 3-8) or located separately and attached to several tanks. A typical lifter roof tank is shown in Figure 3-9. Both types of variable vapor space tanks would be equivalent to a floating roof tank. Variable vapor space tanks are most often used when tank throughput is low because, while breathing losses are virtually eliminated, filling and emptying losses are similar to those in a fixed roof tank.

Pressure tanks provide the greatest degree of control of vapor emissions. For fluids with vapor pressures greater than 11.0 psia and in tanks of 40,000 gallons capacity or higher, pressure tanks or the equivalent are the only acceptable means of reducing vapor losses. Vapor recovery units are generally considered equivalent to pressure tanks.

Regardless of the type of emission control scheme selected, the facility operator must insure compliance with applicable regulations governing the release of toxic vapors. Where the laws specify the use of a particular technology, alternate technologies may be employed provided that the operation can demonstrate the emissions will be equal to or less than emissions which meet the local requirements. Typical controls which may be demonstrated as equivalent to specified tank and roof types are: carbon adsorbers, thermal and catalytic incinerators, and refrigerated condensors. Both carbon adsorbers and condensors allow recovery of volatile organic compounds. Thermal and catalytic incinerators destroy them. If incinerators are used, their emissions must comply with the regulations also.

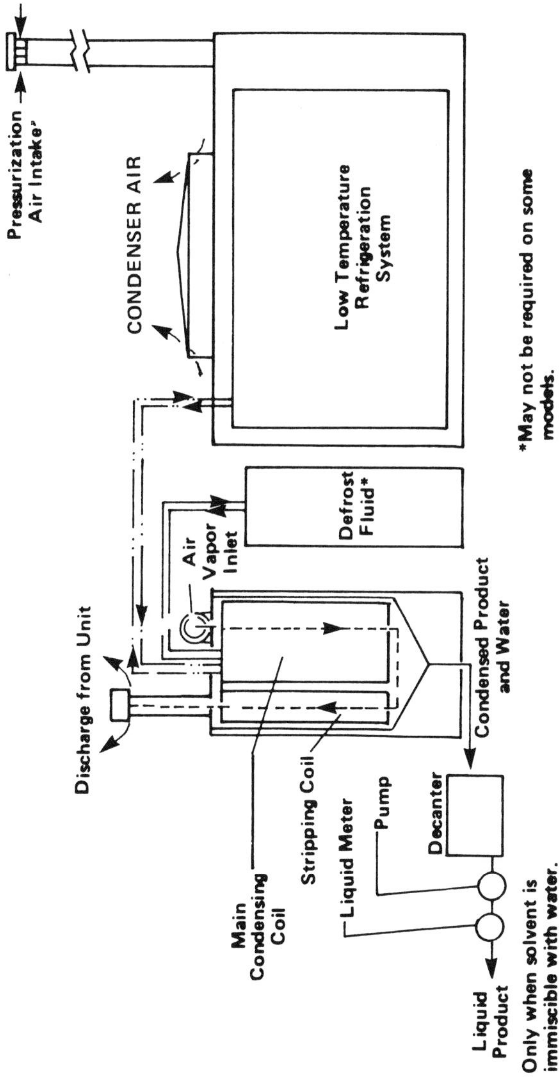

SOURCE: Edwards Engineering Corp.

Figure 3–7 VAPOR RECOVERY UNIT (USING REFRIGERATION)

Table 3-10

EXAMPLES OF RECOVERABLE CHEMICALS

Ketones	Ethers
Acetone	Ethyl ether
Methyl ethyl ketone (MEK)	Tetrahydrofuron
Methyl isobutyl Ketone	Dioxane
Cyclohexanone	
Aromatics	**Hydrocarbons**
Benzene	Hexane
Toluene	Cyclohexane
Xylene	Heptane
Naphthalene	Mineral spirits
Alcohols	**Chlorinated hydrocarbons**
Methanol	Methylene chloride
Ethanol	Methyl chloroform
Isopropanol	Perchloroethylene
Monomers	**Esters**
Vinyl acetate	Vinyl acetate
Acrylic acid	Ethyl acetate
Acrylonitrite	Isopropyl acetate

Source: Edwards Engineering Corp., 1982.

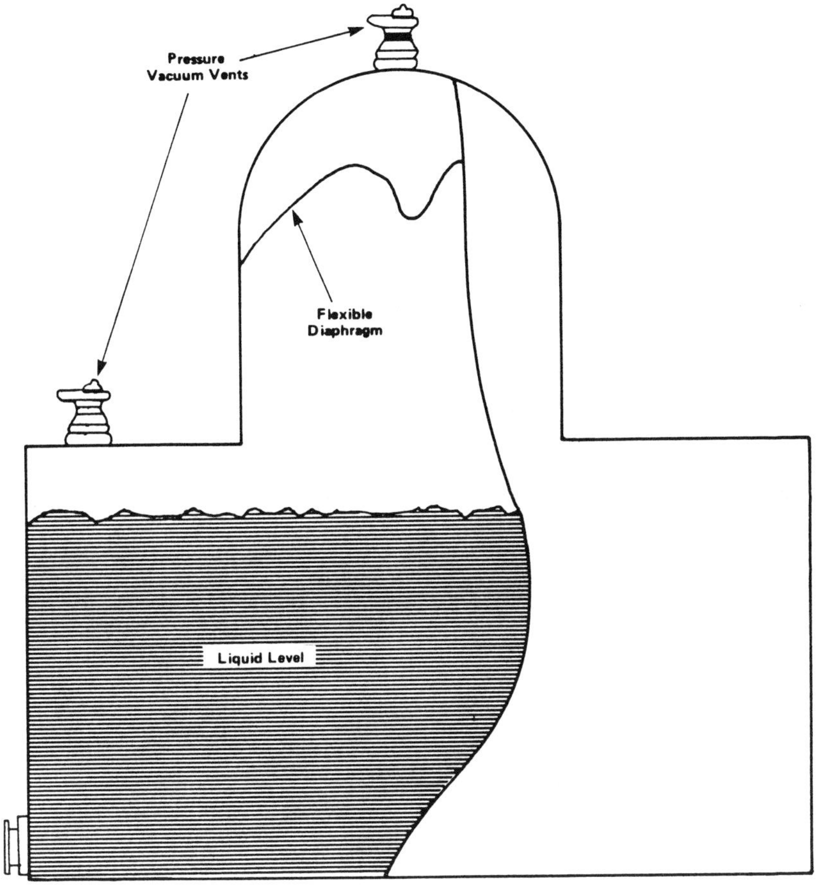

SOURCE: United States Environmental Protection Agency, 1977.

Figure 3–8 FLEXIBLE DIAPHRAGM TANK (INTEGRAL UNIT)

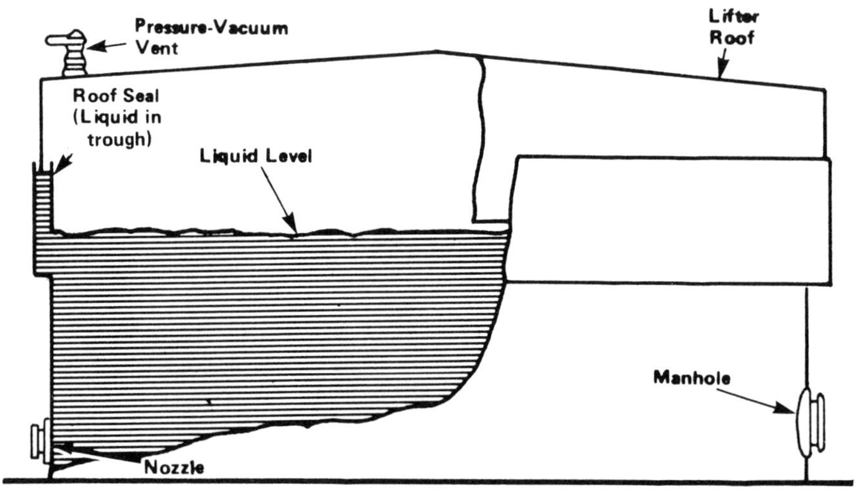

SOURCE: United States Environmental Protection Agency, 1977.

Figure 3–9 LIFTER ROOF STORAGE TANK (WET SEAL)

3.7 SITING CONSIDERATIONS

Constraints on the siting of storage tanks or tank farm facilities arise primarily from the flammability or combustibility of the stored liquids. Toxicity of vapors released in the event of a spill is also of concern in tank siting, but numerous factors specific to the individual chemical storage situation preclude generalized siting guidelines based on toxicity alone. To predict the optimum site, relevant physical, chemical, statistical, and meteorological data should be incorporated into mathematical models specific to the proposed location.

Analysis of the data generated constitutes a hazard and risk analysis of the proposed facility.

Several steps are necessary to perform a hazard and risk assessment during the planning stage of a hazardous materials storage facility. In broad terms, these steps are:

- Identification of the types and causes of potentially hazardous accidents;
- Evaluation of the probability of such accidents occurring at the site; and
- Prediction of the consequences of each accident "scenario" on people and property, particularly off-site.

Each of these steps involves the use of statistical or mathematical models to produce quantitative answers.

The first step, identification, is often aided by a method known as fault tree analysis. This method begins with a hypothetical hazardous condition or accidental release, and proceeds systematically to identify those material, equipment, or human faults which singly or in combination could cause the accident. The logical relationships between the possible causes can then be summarized in a fault tree diagram, permitting identification of reasonable combinations of failures which may lead to a significant hazard.

Step two, evaluation of probabilities, require the collection of failure rate data for each basic fault identified by the fault tree analysis. Such data include hardware component failure rates, material strength information, and historical accident records. By combining these data with the logical events depicted in the fault tree, the probability of the accident can be calculated. Unfortunately, it is often difficult to obtain accurate historical accident statistics, especially involving the particular factors specific to the facility under evaluation. Therefore, probabilities may have to be estimated on the basis of facilities or events similar to the one under study. Further discussion of this may be found in Napadensky and Bodle, 1973.

The third step is to predict the possible consequences of credible accident scenarios. The magnitude of the consequences is determined by factoring into the analysis such variables as: the properties of the hazardous material; the quantity released; meteorological and topographic conditions; extent of downwind dispersion; number and effect of potential ignition sources; etc. Prediction of consequences is necessary for two reasons:

- To establish safe separation distances or "buffer zones" around the facility; and

- To estimate the extent of population and property at risk for comparison with other known and accepted risks.

The possible consequence of greatest importance in siting hazardous materials storage facilities is the release of toxic and flammable gas or vapor clouds.

The prediction of vapor cloud dispersion can be performed on a hand calculator using Gaussian Point Source Model and Pasquill-Gifford disperion coefficients. These are described fully in Turner's *Workbook of Atmospheric Dispersion Estimates*, 1970, and can be applied to models of area sources. Dispersion zones calculated in this manner for many toxic and flammable chemicals are presented in the United States Department of Transportation *Emergency Action Guide for Selected Hazardous Materials*, and are shown in Table 3-11.

Thermal radiation hazards resulting from an ignited vapor cloud can be modeled using the United States Coast Guard Vulnerability Model. A thermal dose probit analysis can be incorporated into the Coast Guard Vulnerability Model to calculate expected numbers of human burn injuries and fatalities. Thermal radiation hazards from pooled fires, such as in the case of liquids spilled in a diked area, can be modeled using the American Gas Association (AGA) computer model developed as part of AGA Project IS-3-1 on LNG Safety Research.

If cloud detonation is of concern, two major effects, air blast overpressure and flying fragments, must be analyzed. Air blast effects are modeled using a "TNT equivalent" approach. This is illustrated in Burgess and Zabetakis' 1973 analysis of a major propane pipeline explosion. Modeling of flying fragments is usually not attempted.

Finally, in order to estimate risk to the population from a hazardous vapor cloud, it is necessary to analyze the surrounding population density in each direction, the presence of ignition sources related to population; and the probability of ignition or toxic effects at each point in the cloud's downwind path.

The estimated risk of property damage can be determined by examining aerial photos or site area maps to locate structures within the calculated hazard radius. This radius can be multiplied by the probability of occurrence of a particular accident scenario and by the estimated value of the property in question. This will result in an estimate of potential property damage liability.

Table 3-11

EVACUATION TABLE FOR SELECTED CHEMICALS

Chemical	*Distance to Evacuate From Immediate Danger Area	*For Maximum Safety, Downwind Evacuation Area Should be:
Acrolein	760 yards	5 miles x 3 miles
Ammonia	90 yards	2,112 feet x 1,584 feet
Chlorine	340 yards	2 miles x 1 1/2 miles
Hydrogen chloride	260 yards	1 1/2 miles x 1 mile
Hydrogen cyanide	130 yards	3,696 feet x 2,112 feet
Hydrogen sulfide	160 yards	1 mile x 1/2 mile
Methyl bromide	50 yards	1,056 feet x 528 feet
Phosgene	820 yards	5 miles x 3 miles
Sulfur trioxide	387 yards	2 miles x 1 mile

*Assuming spill size of 800 square feet, and prevailing wind speed of 6 to 12 mph.

Source: United States Department of Transportation, 1978.

As suggested by the above procedures, modeling and hazard analysis are highly sophisticated procedures which are extremely sensitive to the quality of data utilized, and require knowledgeable interpretation. For these reasons, such analysis should be referred to individuals or organizations familiar with the modeling of situations similar to those to be studied. If modeling or risk analysis is not feasible, siting guidelines based on flammability (such as those of the National Fire Protection Association [NFPA] or the United States Department of Housing and Urban Development [HUD]) are suggested as minimums.

NFPA has established siting guidelines for tank storage of several classes of flammable and combustible liquids (NFPA 30): stable liquids at operating pressures of 2.5 psig or less, stable liquids at operating pressures greater than 2.5 psig, boil-over liquids, unstable liquids, and Class III B liquids. The specific guidelines are a function of tank capacity (see Tables 3-12 through 3-14).

Additional guidance comes from HUD, "Safety Considerations in Siting Housing Projects," 1975, in the form of recommended separation distances between housing developments and storage tanks of liquid industrial fuels and chemicals. These distances consider the threat posed by liquid releases of flammable liquids in terms of the potential for:

- Thermal radiation from a fire causing failure of an adjacent tank and ignition of its contents, and
- Ignition and burning of distant combustible structures or objects.

The distances are a function of the potential spill diameter (D), which in the case of a circular dike is equal to the dike diameter. In the case of a rectangular tank:

$$D = D_{Hydraulic} = \frac{4 \times \text{Area of Dike } (ft^2)}{\text{Perimeter of Dike}}$$

Figure 3-10 contains the HUD safe separation distances for people and cellulosic combustible materials, such as wood.

NFPA also specifies that the minimum in-farm spacing is three feet or 1/6 the sum of the diameters of the two tanks, whichever is greater. If one tank is less than one-half the diameter of the other, then the minimum spacing will be one-half the diameter of the smaller tank.

3.8 SPILL CONTAINMENT AND CONTROL SYSTEMS

Spill containment and control systems refer to actual or planned methodologies for preventing product spilled from storage tanks (primary containment) from adversely impacting human health and the

Table 3-12 LOCATION OF OUTSIDE, ABOVEGROUND LIQUID CHEMICAL STORAGE TANKS

Liquid Type	Type of Tank	Protection	Minimum Distance in Feet from Property Line Which is or can be Built Upon Including the Opposite Side of a Public Way and Shall be no Less than 5 Feet	Minimum Distance in Feet from Nearest Side of Any Public Way or from Nearest Important Building on the Same Property and Shall be not Less than 5 Feet
Stable Liquids (operating pressure 2.5 psig or less)	Floating Roof	Protection for exposures	1/2 times diameter of tank	1/6 times diameter of tank
		None	Diameter of tank but need not exceed 175 feet	1/6 times diameter of tank
	Vertical with Weak Roof to Shell Seam	Approved foam or inerting system on tanks not exceeding 150 feet in diameter	1/2 times diameter of tank	1/6 times diameter of tank
		Protection for exposures	Diameter of tank	1/3 times diameter of tank
		None	2 times diameter of tank but need not exceed 350 feet	
	Horizontal and Vertical with Emergency Relief Venting to Limit Pressures to 2.5 psig	Approved inerting system on the tank or approved foam system on vertical tanks	1/2 times Table 3-12	1/2 times Table 3-12
		Protection for exposures	Table 3-12	Table 3-12
		None	2 times Table 3-12	Table 3-12

Table 3-12 (Cont.)

Liquid Type	Type of Tank	Protection	Minimum Distance in Feet from Property Line Which is or can be Built Upon Including the Opposite Side of a Public Way and Shall be no Less than 5 Feet	Minimum Distance in Feet from Nearest Side of Any Public Way or from Nearest Important Building on the Same Property and Shall be not Less than 5 Feet
Stable Liquids (operating pressure greater than 2.5 psig)	Any Type	Protection for exposures	1 1/2 times Table 3-12 but shall not be less than 25 feet	1 1/2 times Table 3-12 but shall not be less than 25 feet
		None	1 1/2 times Table 3-12, but shall not be less than 50 feet	
Boil-Over Liquids (those liquids with potential to be expelled from a tank during a fire)	Floating Roof	Protection for exposures	1/2 times diameter of tank	1/6 times diameter of tank
		None	Diameter of tank	1/6 times diameter of tank
	Fixed Roof	Approved foam or inerting system	Diameter of tank	1/3 times diameter of tank
		Protection for exposures	2 times diameter of tank	2/3 times diameter of tank
		None	4 times diameter of tank but need not exceed 350 feet	2/3 times diameter of tank

Table 3-12 (Cont.)

Liquid Type	Type of Tank	Protection	Minimum Distance in Feet from Property Line Which is or can be Built Upon Including the Opposite Side of a Public Way and Shall be no Less than 5 Feet	Minimum Distance in Feet from Nearest Side of Any Public Way or from Nearest Important Building on the Same Property and Shall be not Less than 5 Feet
Unstable Liquids (those which will vigorously polymerize, decompose, condense, or react under conditions of shock, pressure, or temperature)	Horizontal and Vertical Tanks with Emergency Relief Venting to Permit Pressure Not in Excess of 2.5 psig	Tank protected with any one of the following: approved water spray, approved inerting, approved insultation and refrigeration, approved barricade	Table 3-12 but not less than 25 feet	Not less than 25
		Protection for exposures	2 1/2 times Table 3-12 but not less than 50 feet	Not less than 50 feet
		None	5 times Table 3-12 but not less than 100 feet	Not less than 100 feet
	Horizontal and Vertical Tanks with Emergency Relief Venting to Permit Pressure Over 2.5 psig	Tank protected with any one of the following: approved water spray, approved inerting, approved insulation and refrigeration, approved barricade	2 times Table 3-12 but not less than 50 feet	Not less than 50 feet
		Protection for exposures	4 times Table 3-12 but not less than 100 feet	Not less than 100 feet
		None	8 times Table 3-12 but not less than 150 feet	Not less than 150 feet

Source: NFPA 30.

Table 3-13

REFERENCE TABLE FOR
USE WITH TABLE 3-12

Capacity Tank (Gallons)	Minimum Distance in Feet from Property Line Which Is or Can Be Built Upon, Including the Opposite Side of a Public Way	Minimum Distance in Feet from Nearest Side of Any Public Way or from Nearest Important Building on the Same Property
275 or less	5	5
276 to 750	10	5
751 to 12,000	15	5
12,001 to 30,000	20	5
30,001 to 50,000	30	10
50,001 to 100,000	50	15
100,001 to 500,000	80	25
500,001 to 1,000,000	100	35
1,000,001 to 2,000,000	135	45
2,000,001 to 3,000,000	165	55
3,000,001 or more	175	60

Source: NFPA 30.

Table 3-14

LOCATION OF ABOVEGROUND TANKS STORING
CLASS III B LIQUIDS

Capacity (Gallons)	Minimum Distance in Feet from Property Line Which Is or Can Be Built Upon, Including the Opposite Side of a Public Way	Minimum Distance in Feet from Nearest Side of Any Public Way or from Nearest Important Building on the Same Property
12,000 or less	5	5
12,001 to 30,000	10	5
30,001 to 50,000	10	10
50,001 to 100,000	15	10
100,001 or more	15	15

Source: NFPA 30.

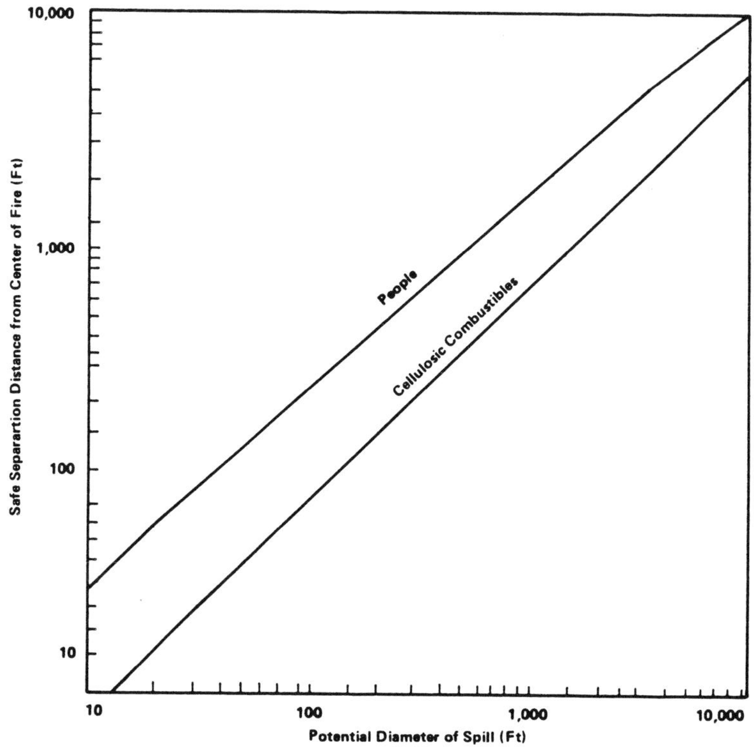

SOURCE: United States Department of Housing and Urban Development, 1975.

Figure 3–10 SAFE SEPARATION DISTANCES FROM SPILLS OF COMMON LIQUID INDUSTRIAL FUELS-FIRE THREAT

environment. Physical systems to contain the spread of spilled product are considered preferable to "contingency" plans, in that they are preventive rather than reactive in nature. Every effort should be made to ensure that hazardous substances are not allowed to run off the facility property, whether by surface drainage, or in sewer systems or groundwater.

Attention to certain design considerations can assist in achieving more effective drainage and spill controls:

- Drainage lines should be laid out so that areas of relatively frequent, yet light product spillage, e.g. loading racks or pump and equipment areas, may drain by gravity to an oil-water separator;

- A positive contingency control plan should be provided for areas of rare, yet potentially heavy, product spillage, e.g., the tank farm area, with provision for drainage to treatment facilities at a later time;

- Controlled drainage of storm runoff from driveway and other areas subject to oil contamination should be provided via a sewer or other conveyance network to the separator;

- Retention or diversion of potential spills and leaks from piping and equipment systems should be provided for in such a manner that they can be cleaned up within a localized area; and

- Pure product flows (i.e., equipment drain down, blow-offs, flushing) should be segregated from other drainage and runoff, and directed to collection tanks to alleviate excessive separator loading.

The flowchart in Figure 3-11 indicates the necessary elements in a spill containment and control system.

3.8.1 Types of Containment

An analysis of bulk plants reveals the following causes of spills in order of significance:

- Overfilling of tanks;

- Leaking tanks, pipes, pumps, and other equipment; and

- Spills during tank truck or tank car loading and unloading and marine transfers.

Development of appropriate spill containment measures should include review of potentially susceptible areas. Contaminated waters from outside areas should be prevented from entering facility property, where feasible. This may be accomplished by providing low berms, curbs on paved areas, interceptor ditches on open land, or diversion of natural drainage. All ditches, sewers, and natural slopes where

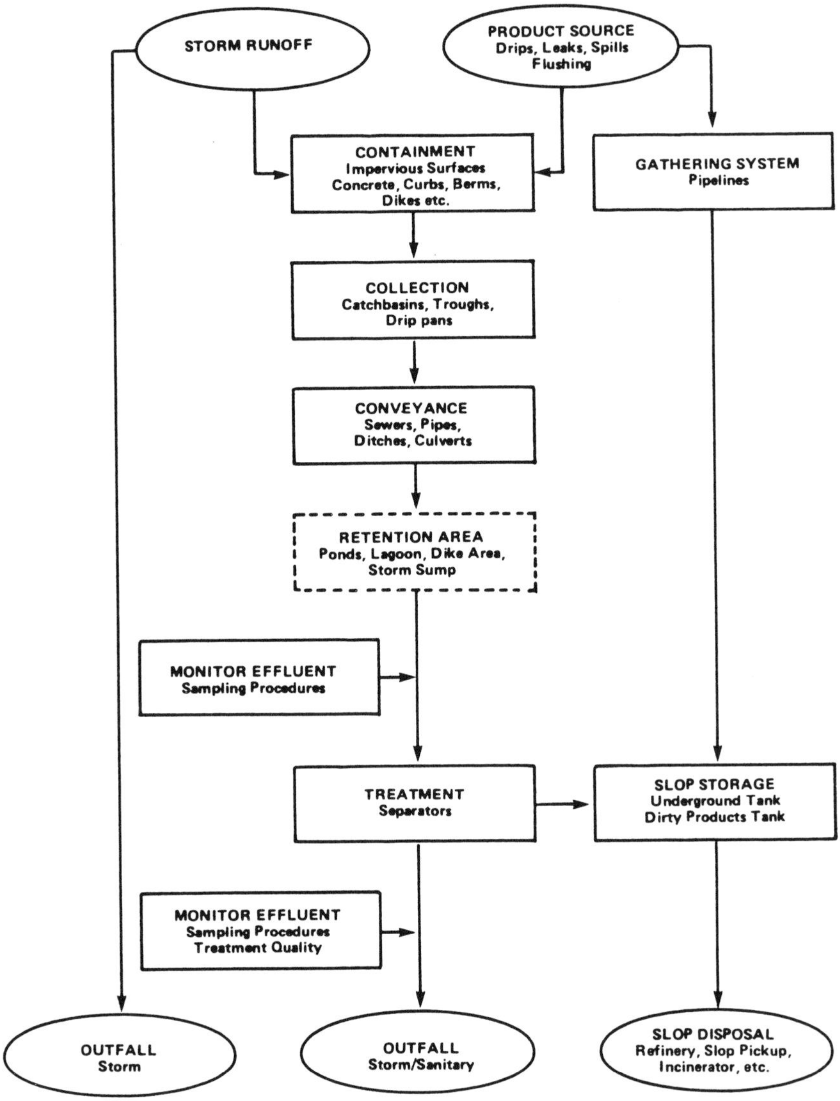

SOURCE: PACE, 1980.

Figure 3–11 SPILL CONTAINMENT AND CONTROL FLOW

product may escape with runoff leaving the property should be checked. This should include a review of a topographic survey made of the site and surrounding area as well as a physical inspection of the location. Such means as necessary to insure against any product runoff from a major break or spill should be provided.

Upon review of the potential spill sources, the appropriate containment, collection, conveyance, and retention measures may be selected. The containment system is an arrangement of impervious surfaces (concrete, asphalt, membrane, etc.) surrounded by curbs, gutters, dikes, etc. The purpose is to prevent any flow from leaving the immediate area. The collection system is a series of components that collect pure product and potential chemical-bearing flow. These may be collection troughs, drainage pans, funnels, catch basins, etc.

The conveyance network is a system of pipes, channels, sewers, culverts, etc., necessary to transport flow from the collection points to slop tanks, retention areas, treatment facilities, or outfall, dependent upon flow characteristics. The conveyance network will collect similar types of flow from different areas and route the flows to a common destination. Care must be taken that the conveyance network is constructed of suitably resistant materials.

If a large flow volume occurs at a facility, retention facilities may be needed. A retention facility may be a pond, lagoon, dike area, or storm pump. The purpose of the retention facility is to temporarily store potential chemical-bearing flow to allow a controlled rate of input into the treatment facilities.

Spills can be more effectively controlled at loading and unloading areas by providing impervious surfaces to the loading and unloading areas. An impervious surface would typically be provided with a peripheral border or other means to trap all spills or rain within its confines and route the liquid to treatment areas. Sufficient capacity should be provided to contain or hold back the contingency spill volume.

3.8.2 Material Selection

Material employed in the construction of dikes or retaining walls should be compatible with the material being stored. For example, the use of concrete blocks, which are frequently used to construct retaining walls, should be avoided in the presence of hydrochloric, hydrofluoric, muriatic, nitric, sulfuric, and sulfurous acid, as well as nitrates and sulfates of ammonia.

Materials useful for spill containment and collection surfaces include natural permeable and impermeable soils, synthetic membrane liners, soil additives, cement, and asphalt. Choice of the appropriate surface material is dependent upon the following considerations.

- The degree of impermeability required;
- The extent of longevity or weather resistance required;
- The compatibility with the material being stored.

Some of the commonly used containment surface materials and their applications are as follows:

- Natural permeable soils, generally used only for cover protection of impermeable surfaces and as clay or membrane liners.

- Natural clays, commonly used as a relatively inexpensive impermeable surface where minor leaching of the stored substance is not critical.

- Treated bentonite clays provide low degrees of permeability at moderate cost.

- Synthetic membrane liners provide high degrees of impermeability and chemical resistance, but require extensive surface preparation.

- Asphalt is widely used for containment of aqueous solutions, but may be inadequate for some hydrocarbons.

- Concrete is a durable, but somewhat permeable, surface material.

3.8.3 Design Capacity

The capacity of a diked area or retention pond should be sufficient to contain the entire volume of the largest tank feeding that area or pond, plus a minimum allowance of 10% to accommodate accumulated precipitation or other water. Depending on tank volume, additional freeboard may be required to contain the surge and waves resulting from a sudden, rapid tank failure. The capacity of a diked area is calculated by multiplying the surface area within the dike (less the "floorspace" area physically occupied by tanks) by the height of the dike or wall. This is illustrated in Figure 3-12.

Example of Calculation of Required Capacity

- For Figure 3-12, determine height of walled area (h), so that the capacity (C) is adequate. The desired capacity is the volume of the largest tank, 20,000 gallons, plus 10%.

$$\begin{aligned} C &= 20{,}000 + (.1)\ (20{,}000) \\ &= 22{,}000 \text{ gal} \\ &= (22{,}000 \text{ gal})\ (.1337 \text{ ft}^3/\text{gal}) \\ &= 2{,}942 \text{ ft}^3 \end{aligned}$$

- The available surface area (A) is the walled area less the cross sectional area (πr^2) of the two vertical tanks (the two horizontal tanks are elevated on saddle supports which do not appreciably reduce the available surface area.)

$$\begin{aligned} A &= (75 \times 30) - 2\ (\pi)\ (10/2)^2 \\ &= 2{,}093 \text{ ft}^3 \end{aligned}$$

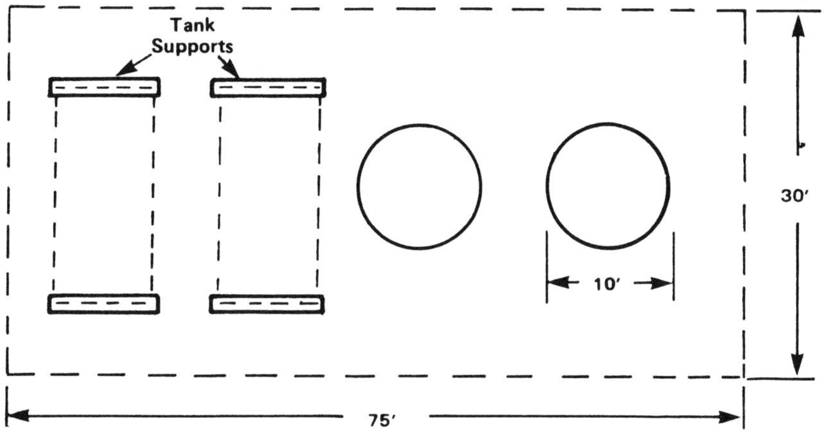

CONTAINMENT AREA PLAN VIEW

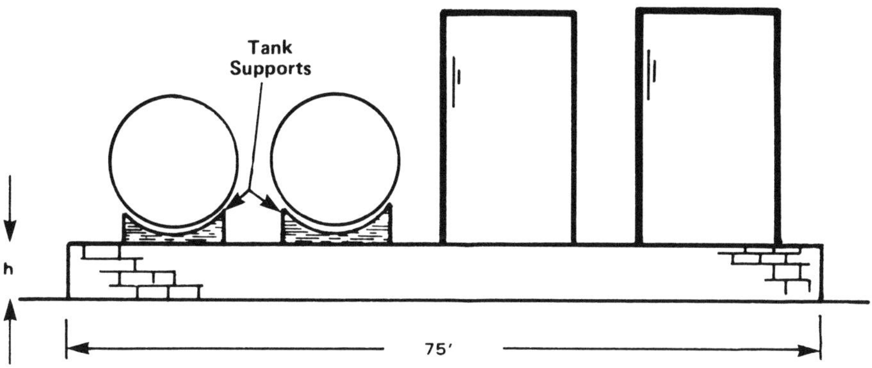

CONTAINMENT AREA ELEVATION

SOURCE: Ecology and Environment, Inc., 1983.

Figure 3–12 ILLUSTRATIVE METHOD FOR DETERMING CONTAIMENT AREA CAPACITY (NOT TO SCALE)

- The available surface area (A) is the walled area less the cross sectional area (πr^2) of the two vertical tanks (the two horizontal tanks are elevated on saddle supports which do not appreciably reduce the available surface area.)

  $$A = (75 \times 30) - 2\,(\pi)\,(10/2)^2$$
  $$= 2{,}093 \text{ ft}^3$$

- The required height (h) equals the required capacity (C) divided by the available area (A).

  $$h = C/A = 2942 \text{ ft}^3/2093 \text{ ft}^3$$
  $$= 1.4 \text{ ft}$$
  $$= 17 \text{ in}$$

Although technically the cross sectional area of a vertical tank that fails does not reduce the available surface area, the practice of including all such tanks in the calculation provides an additional margin of safety in the computed capacity.

In the case of retention ponds, the required capacity should equal the volume of the largest tank that could potentially drain into it, plus a 10% freeboard allowance. Typical earthen dike construction is illustrated in Figures 3-13 and 3-14.

3.8.4 Drainage Collection

Diked areas and retention ponds should be equipped with release valves to permit the drainage of accumulated precipitation or runoff from these areas. These valves should be of the manual, open and close variety, not flapper models or automated systems. The valves should be chained and locked in the closed position when not in use. In this way, unwarranted discharges of contained product can be avoided. Legitimate releases of accumulated rainwater should be monitored to insure that this effluent is not contaminated with product, and should be included in a regular maintenance schedule. Fill or drain pipes should not protrude through a man-made wall or earthen berm. This would provide a mechanism by which a liquid could breach the structure, and so should be avoided.

Drainage from facilities which store "lighter than water" products should be gravity-fed to a separator. The separator should be pumped out on a regular basis to prevent overflow. Likewise, drainage from chemical storage areas that is suspected of being contaminated should be treated by appropriate technology.

3.9 IGNITION SAFEGUARD

Hazardous materials storage tanks are subject to many potential sources of ignition. These include open flames, lightning, smoking, cutting and welding, hot surfaces, friction, sparks, spontaneous ignition, chemical reaction, and radiant heat. Safeguards are imperative in situations where flammable vapor-air mixtures could be ignited.

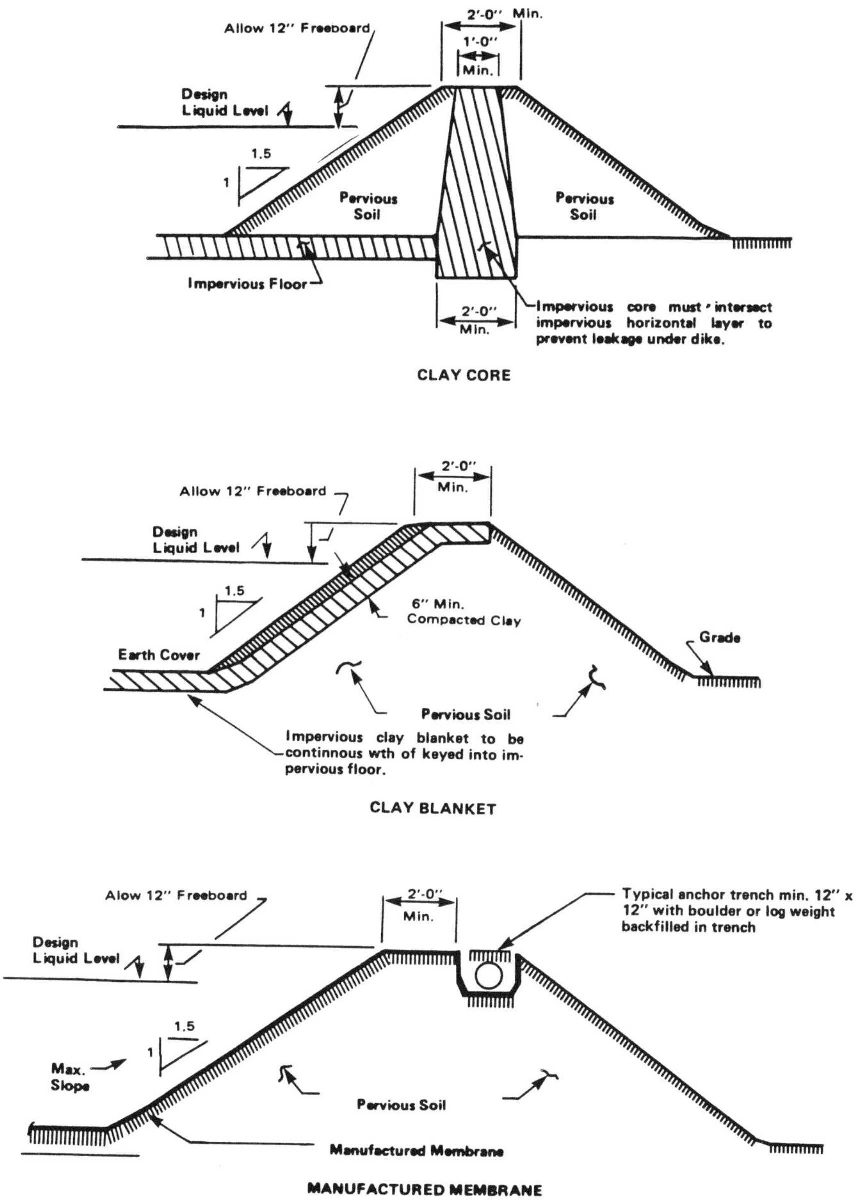

SOURCE: PACE, 1980.

Figure 3—13 TYPICAL EARTHEN DIKES

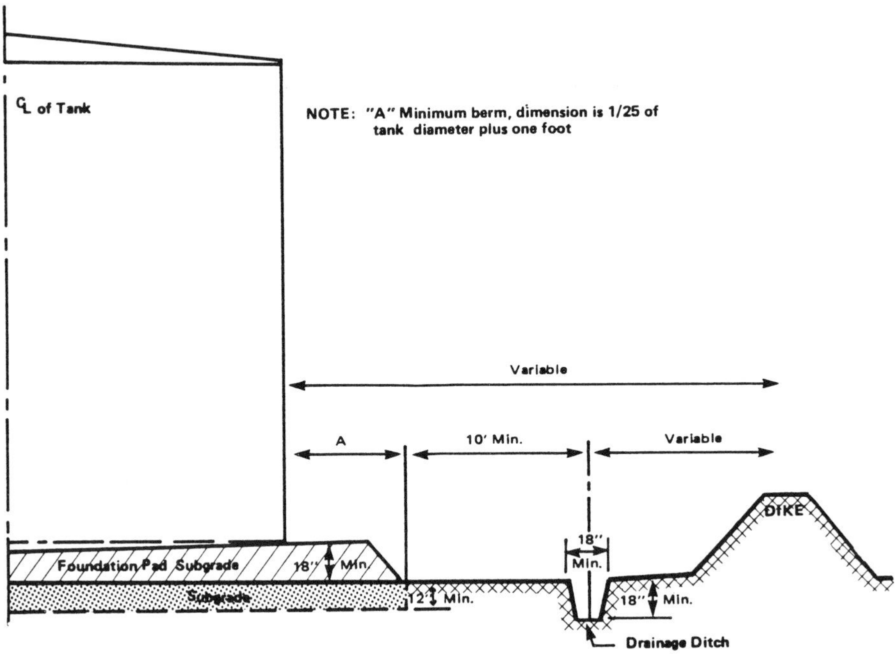

SOURCE: PACE, 1980.

Figure 3–14 TANK LOT GRADING FOR SPILL CONTROL

The extent of fire protection shall be determined by an evaluation based upon sound fire protection engineering principles, analysis of local conditions, hazards within the facility, and exposure to or from other property. The evaluation should determine, as a minimum:

- The type, quantity, and location of equipment necessary for the detection and control of fires, leaks, and spills;
- The methods necessary for protection of equipment and structures from the effects of fire exposure;
- Fire protection water systems;
- Fire extinguishing and control equipment;
- The equipment and processes to be incorporated within an Emergency Shutdown System (ESS), including an analysis of subsystems, if any, and the need for depressuring specific vessels or equipment during a fire;
- The type and location of sensors necessary to initiate automatic operation of the ESS or its subsystems;
- The availability and duties of individual plant personnel and the availability of external response personnel during an emergency; and
- The protective equipment and special training needed by the individual plant personnel for their respective emergency duties.

Table 3-15 provides information on various techniques which can be used to prevent fires associated with the storage of selected hazardous materials.

3.9.1 Ignition Control

Under normal conditions, static, electrical, or mechanical sparks are the most likely sources of ignition of stored flammable and combustible materials. Electrical sparks can be prevented by using intrinsically safe electrical equipment in the vicinity of storage tanks. Guidelines for safe electrified equipment are detailed in NFPA 70: National Electrical Code. Prevention of mechanical sparks can be achieved by avoiding intermittent metal-to-metal contact and other spark-causing situations. Static sparks can result as liquid moves around in a tank and accumulates static charge. To minimize the possibility of static ignition, the following safeguards should be implemented:

- "Splash-filling" of tanks or other strenuous agitation of contents should be avoided. The discharge from a fill hose into a tank should be close to the bottom of the tank.
- The velocity of the incoming stream should be limited to 1 m/sec. to minimize the agitation of tank contents and subsequent build-up of charge.

Table 3-15

FIRE PROTECTION TECHNIQUES FOR STORAGE OF SELECTED HAZARDOUS MATERIALS

Chemical	Fire Prevention
Cyanides	Avoid physical damage; insulate from acids
Chromic acid	Separate from oxidizable materials; avoid storage on wooden surfaces; remove spills
Hydrofluoric acid	Corrodes many materials except lead, wax, polyethylene, and platinum; store in vented area
Hydrochloric acid	Separate from oxidizable materials; store in cool, vented area; avoid contact with common metals
Nitric acid	Separate from metallic powders, carbides, hydrogen sulfide, turpentine, organic acids, and oxidizable materials; avoid direct sunlight
Sulfuric acid	Avoid nitrates, powdered metals, chlorates, and other oxidizable materials
Acetic acid	Avoid oxidizable and combustible materials; keep above freezing point
Ferric chloride	Protect against physical damage; store in cool, vented area
Ammonium persulfate	Keep away from strong oxidizers like chlorates, nitrates, and nitrites
Caustics	Store in dry place; avoid moisture; separate from ignitable materials
Ammonia	Store in cool, vented area; avoid combustible materials; avoid chlorine, bromine, iodine, acids
Alkaline wastes	Store in cool, vented area; avoid flammable materials
Mercury	Store in cool, vented area away from combustibles
Tetraethyl lead and lead oxide mixed	Store in cool, vented area; avoid strong oxidizing agents; store in sprinklered area
Lead compounds and oxides	Store in cool, dry place; avoid storage on wood floors; avoid combustibles
Zinc compounds	Store in cool, dry, vented area; avoid strong acids and alkalies
Sodium compounds	Store away from combustibles; avoid high temperatures
Aluminum, phosphorus compounds, and sulphur compounds	Keep dry; insulate from acids, caustics, and chlorinated hydrocarbons; avoid combustible materials

Source: Ecology and Environment, Inc., 1982.

- Ungrounded objects should be eliminated.
- Spark promoters within a tank, such as protruding metal surfaces or floating objects, should be eliminated.
- On RP tanks, all metallic objects should be bonded together and grounded.
- Tank fill nozzles should be electrically grounded to the tank during tank loading.

Ignition due to a direct lightning strike is not preventable. The possibility of ignition from lightning can be minimized, however, by providing for the dissipation of charge, through adequate grounding. Tanks with fixed metallic roofs and horizontal tanks are generally adequately protected, since all metallic components are in contact with each other. In other situations the following should be considered:

- Non-conducting roofs should be provided with a metal covering that is in contact with the conductive shell, lightning rods, conducting masts, or overhead ground wires.
- Metallic shunts (straps) should be provided between metallic floating roofs and the metallic shell to overcome the insulating effect of the rubber seal between the two.

The propagation of open flames into or out of a tank can be prevented with the use of conservation valves or flame arrestors. On a facility-wide basis, flame propagation can be reduced by maintaining adequate supplies of firefighting materials. Water spray, foams, inert gases, and dry chemicals are all used for firefighting. Appropriate firefighting materials for common hazardous chemicals are specified in the NFPA Fire Protection Guide on Hazardous Materials.

The possibility of ignition due to radiant heat can be minimized by painting storage tanks that are to contain flammable and combustible materials with reflective paints. Aluminum-based paints are commonly used for this application. Reflective paints are also used to some extent to control vapor emissions.

The ignition hazards of cutting and welding operations are most easily eliminated by only performing such hot work after vapor-space testing has clearly shown that flammable vapor/air mixtures do not exist in or near the work area. Additionally, cigarette smoking should be permitted only in designated and properly posted areas. Vehicles, small engines, and other mobile equipment, which are not rated as "intrinsically safe" or "explosion-proof" should be prohibited near flammable material storage or impounding areas, except when specifically authorized and supervised.

Additional information on ignition safeguards and fire prevention related to chemical storage can be found in the following codes:

- NFPA 77, Recommended Practice on Static Electricity;
- NFPA 78, Lightning Protection Code;
- API-RP2003, Protection Against Ignitions Arising out of Static, Lightning, and Stray Currents;
- NFPA 70, National Electrical Code;
- NFPA Fire Protection Guide on Hazardous Materials; and
- NFPA 30 Flammable Liquids Code.

3.9.2 Fire and Leak Control

Areas such as enclosed buildings which have a potential for flammable gas concentrations, should be appropriately monitored.

Continuously monitored temperature sensors or flammable gas detection systems should be provided to sound an alarm at the plant site and at a continuously attended location, if the plant site itself is not continuously attended. To provide an adequate margin of safety in flammable gas detection systems, it is recommended that the alarm be set to sound at not more than 25 percent of the lower flammable limit of the gas or vapor being monitored.

Fire detectors should be installed to sound an alarm at the plant site, and at a continuously attended location if the plant site itself is not continuously attended. In addition, fire detectors may activate appropriate portions of the ESS.

The appropriate detection systems should be designed, installed, and maintained in accordance with the following applicable NFPA standards:

- No. 72A, Installation, Maintenance, and Use of Local Protective Signaling Systems;
- No. 72B, Installation, Maintenance, and Use of Auxiliary Protective Signaling Systems;
- No. 72C, Installation, Maintenance, and Use of Remote Station Protective Signaling Systems;
- No. 72D, Installation, Maintenance, and Use of Proprietary Protective Signaling Systems;
- No. 72E, Automatic Fire Detectors; and
- No. 1221, Installation, Maintenance, and Use of Public Fire Service Communications.

3.9.3 Fire Extinguishing and Control

Portable fire extinguishers suitable for the material stored should be available at strategic locations within a hazardous materials storage facility and on tank vehicles. These extinguishers should be provided and maintained in accordance with NFPA Standard No. 10 (Portable Fire Extinguishers). Fixed fire extinguishers and other fire control systems also may be appropriate. If provided, such systems shall be designed, installed, and maintained in accordance with the following applicable NFPA standards:

- No. 11, Foam Extinguishing Systems;
- No. 11A, High Expansion Foam Systems;
- No. 11B, Synthetic Foam and Combined Agent Systems;
- No. 12, Carbon Dioxide Extinguishing Systems;
- No. 12A, Halogenated Fire Extinguishing Agent Systems--Halon 1301;
- No. 12B, Halogenated Extinguishing Agent Systems--Halon 1211;
- No. 16, Installation of Foam-Water Sprinkler Systems and Foam-Water Spray Systems; and
- No. 17, Dry Chemical Extinguishing Systems.

Selected methods for extinguishing specific chemical fires are described in Table 3-16. Further guidance may be obtained from NFPA Standard Number 10, and NFPA Publication 325M, Fire Hazard Properties of Flammable Liquids, Gases, and Volatile Solids (also found in Fire Protection Guide on Hazardous Materials).

3.10 FAIL-SAFE AND WARNING DEVICES

Accidental spills of hazardous substances can occur for a number of reasons. These may include spills during product transfer, overflow during filling operations, and leakage or rupture of tanks or appurtenances because of corrosion vapor build-up or other circumstances. Spills due to such causes can be prevented, or their effects minimized, through application of appropriate sensing and warning devices. Such devices may activate shut-off or diversion mechanisms, or may simply provide audible or visual alarms of adverse circumstances. The devices can be classified as level detectors, leak monitors, and gas detectors.

Level Detection devices are the main component of overfill prevention systems. They use level sensors and gauges to detect liquid levels in the tank, and activate alarms to warn of potential overfill situations. These may be linked to electronic or mechanical devices to shut down filling operations or divert flow to emergency overflow tanks. Types of level detection devices are:

Table 3-16

SELECTED METHODS OF EXTINGUISHING CHEMICAL FIRES

Chemical	Extinguishing Methods
Cyanides	Use water; do not use CO_2 extinguishers; avoid toxic fumes
Chromic acid	Use water; caution should be exercised against possibility of stream explosion
Hydrofluoric acid	Use water; neutralize with soda ash or lime; if water is ineffective, use "alcohol foam"
Hydrochloric acid	Use water; neutralize with soda ash or slaked lime
Nitric acid	Use a water spray; neutralize with soda ash or lime
Sulfuric acid	Use large amounts of water; reaction may occur; neutralize with ash or lime; sand or gravel also will help
Acetic acid	Use water spray, dry chemical, "alcohol foam," or carbon dioxide
Ferric chloride	Use water
Ammonium persulfate	Use water spray or water flooding; avoid toxic fumes
Caustics	Flood with water; avoid spattering or splashing
Ammonia	Stop flow of material; use water to keep container cool; avoid fumes
Alkaline wastes	Use water; neutralize with dilute acid (acetic) if necessary
Mercury	Use water; avoid toxic mercury vapor
Tetraethyl lead and lead oxide mixed	Fight fires from explosion-restraint location; use water, dry chemical, foam, or carbon dioxide
Lead compounds and oxides	Use flooding amounts of water
Zinc compounds	Smother with suitable dry powder
Sodium compounds	Use water, dry powder; neutralize with appropriate chemical, if necessary
Aluminum, phosphorus compounds, and sulphur compounds	Do not use water; smother with suitable dry powder

Source: Ecology and Environment, Inc., 1982.

- Float-actuated devices,
- Displacement systems,
- Electrical capacitance sensors,
- Optical sensors,
- Ultrasonic sensors,
- Thermal conductivity sensors, and
- Pressure sensors.

Leak detection devices are used to monitor leaks onto the ground surface or into the groundwater table surrounding both underground and aboveground tanks. These include electrical resistivity sensors, interstitial fluid or pressure sensors for double-walled tanks, and thermal conductivity sensors.

Leak detection devices can be placed within a diked area to quickly reveal the first occurrence of a spill. For underground tanks, sensors can be placed within slotted groundwater monitoring wells situated close to the storage area in a downgradient direction. Spilled material that has saturated the soil and entered the groundwater table will be readily detected. However, the variety of chemicals that can be detected in this manner may be limited by specific physical or chemical properties. Table 3-17 lists chemicals detectable with thermal conductivity-type sensors. Equipment manufacturers should be contacted directly for information on specific applications.

Gas detectors can be used to monitor a wide variety of flammable, nonflammable, and toxic gases and vapors. These devices are available as permanent installations, or as portable instruments for tracking the source, direction, and intensity of a gas or vapor leak. They may also be used to activate audible or visual alarm systems, ventilation equipment, or process interruption equipment.

Combustible gas detectors can be used to detect conditions which present an explosive hazard as a result of the release of any flammable or combustible gases, or vapors from flammable or combustible liquids. Other detectors are available for specific gases, such as carbon monoxide or hydrogen sulfide. Infrared analyzers are particularly useful in identifying single compounds that are infrared-active. These would include carbon dioxide, halogenated hydrocarbons, and most other hydrocarbons.

At unattended hazardous materials storage facilities, the appropriate sensing and warning devices should be connected to an alarm circuit. This circuit should transmit the alarm to a continuously attended facility to indicate any symptoms of trouble such as abnormal temperatures, pressure increases, level changes, etc.

Table 3-17

COMMON CHEMICALS DETECTABLE WITH THERMAL CONDUCTIVITY SENSORS

Acetaldehyde	Ethylene Glycol Monobutyl Ether
Allyl Alcohol	Gasoline
Aniline	Glycerine
Benzene	Isoprene
Benzyl Chloride	Kerosene
Butyl Acetate	Methanol
Carbon Disulfide	Methyl Isobutyl Ketonc
Carbon Tetrachloride	Monoethanolamine
Chlorobenzene	Naphtha
Chloroform	Nitrobenzene
Cresol	Nitropropane
Cumene	Nitrotoluene
Cyclohexane	Phenol
Cyclohexanone	Polychlorinated Biphenyls
Cyclohexanol	Styrene
Dichloromethane	Tetrachloroethane
Diethylamine	Tetrachloroethylene
Epichlorohydrin	Tetrachloromethane
Ethanol	Toluene
Ethyl Acetate	Trichloroethylene
Ethyl Benzene	Vinylidene Chloride
Ethylene Glycol	Xylene

Source: Mallory Components Group, 1982.

BIBLIOGRAPHY

Air Pollution Control Association 1971, Control of Atmospheric Emissions From Petroleum Storage Tanks, JAPCA.

American Concrete Institute, Design and Construction of Circular Prestressed Concrete Structures, No. 344, Detroit, MI.

American Gas Association, July 1, 1974, Project IS-3-1, LNG Safety Program, Interim Report on Phase II Work, Arlington, VA.

American Petroleum Institute (API), 1977, API Specification for Shop Welded Tanks for Storage of Production Liquids, API Spec. 12F, American Petroleum Institute Publishers, Washington, D.C.

_________, 1980, Welded Steel Tanks for Oil Storage, API Standard 650, American Petroleum Institute Publishers, Washington, D.C.

_________, 1978, Recommended Rules for Design and Construction of Large, Welded, Low-Pressure Storage Tanks, API Standard 620, American Petroleum Institute Publishers, Washington, D.C.

_________, 1982, Protection Against Ignitions Arising Out of Static, Lightning, and Stray Currents, API Recommended Practice 2003, American Petroleum Institute Publishers, Washington, D.C.

_________, 1982, Venting Atmospheric and Low-Pressure Storage Tanks, API Standard 2000, American Petroleum Institute Publishers, Washington, D.C.

_________, 1976, Recommended Practice for Design and Installation of Pressure Relieving Systems in Refineries, API Recommended Practice 520, Parts I and II, American Petroleum Institute Publishers, Washington, D.C.

_________, 1969, Guide for Pressure Relief and Depressing Systems, API Recommended Practice 521, American Petroleum Institute Publishers, Washington, D.C.

_________, 1969, Flanged Steel Safety Relief Valves, API Standard 526, Second edition, American Petroleum Institute Publishers, Washington, D.C.

_________, 1979, Installation of Underground Petroleum Storage Systems, API Publication 1615, American Petroleum Institute Publishers, Washington, D.C.

American Society of Mechanical Engineers, 1974, Section II - Material Specifications, ASME Boiler and Pressure Vessel Code, McGraw-Hill Publications, New York, NY.

________, 1974, Section VIII - Pressure Vessels - Division Land, ASME Boiler and Pressure Vessel Code, McGraw-Hill Publications, New York, NY.

________, 1974, Section X - Fiberglass Reinforced Plastic Pressure Vessels, ASME Boiler and Pressure Vessel Code, McGraw-Hill Publications, New York, NY.

American Society of Testing Materials, Specification for Fiberglass Reinforced Polyester Tanks, ASTM Specification D3299, Philadelphia, PA.

American Water Works Association, 1973, Welded Steel Elevated Tanks, Standpipes, and Resevoirs for Water Storage, AWWA Standard D100, American Water Works Associations, Denver, CO.

British Valve Manufacturers Association, 1966, Valves For The Control of Fluids, Pergamon Press, Oxford.

Burgess, D.S., and M.G. Zabetakis, 1973, Detonation of a Flammable Cloud Following a Propane Pipeline Break, United States Bureau of Mines Report of Investigations, RI 7752, Pittsburgh, PA.

Burk, H., 1981, Conceptual Design of Refinery Tankage, Chemical Engineering, 88 (17): 107-110.

Clarke, J., 1971, Recent Tank Bottom and Foundation Problems, American Petroleum Institute - Division of Refining, Dallas, TX.

Cotz, V., 1972, Plant Engineer's Manual and Guide, Prentice-Hall, Englewood Cliffs, NJ.

Edwards Engineering Corp., 1982, Manufacturers literature, Pompton Plains, NJ.

Eisenbuerg, N.A., C.J. Lynch, and R.J. Breeding, 1975, Vulnerability Model: A Simulation System for Assessing Damage Resulting from Marine Spills, Report No. CG-D-136-75, United States Coast Guard Office Of Research and Development, Washington, DC.

Epstein, L.D., 1981, Costs of Standard Vertical Storage Tanks and Reactors, Chemical Engineering, 88(14): 141-142.

Factory Mutual, Handbook of Industrial Loss Prevention, McGraw-Hill Publications, New York, NY.

Hepner, I., ed., 1962, Materials of Construction for Chemical Plant, Leonard Hill Ltd., London.

Higdon, A., E. Ohlsen, W. Stiles, J. Weese, and W. Riley, 1976, *Mechanics of Materials*, 3rd Ed., John Wiley and Sons, New York, New York.

Holzhaver, R., 1980, Underground Liquid Fuel Storage Tanks, *Plant Engineering*, 34(8):98-105.

Hutchinson, J., 1976, *ISA Handbook of Control Valves*, Second Edition, Instrument Society of America, Triangle Park, NC.

Kraus, N., 1980, FRP Underground Horizontal Tanks for Corrosive Chemicals, *Chemical Engineering*, 87(3):125-28, New York, NY.

Mallinson, J., 1969, *Chemical Plant Design With Reinforced Plastics*, McGraw-Hill Publications, New York, NY.

Malloy Components Group, 1982, "Pollutant Systems Detectability Listings," Indianapolis, IN.

Marshall, W.W., 1981, Construction Materials for Chemical Process Industries, *Chemical Engineering*, No. 368 (May 1981): 221-228.

Napadensky, H.S. and Bodle, W.W., 1973, Safe Siting of Petrochemical Plants, presented at the National Symposium on Occupational Safety and Health, sponsored by the American Chemical Society, June 4-6, 1973, Washington, DC.

National Fire Protection Association (NFPA), 1978, *Fire Protection Guide on Hazardous Materials*, National Fire Protection Association, Boston, MA.

———, 1981, *Fire Protection Handbook, Fifteenth Edition*, National Fire Protection Association, Boston, MA.

———, 1981, *Flammable and Combustible Liquids Code, National Fire Code*, Section 30, Volume 3, National Fire Protection Association, Boston, MA.

———, 1981, *Hazardous Chemical Data, National Fire Code*, Section 49, Volume 13, National Fire Protection Association, Boston, MA.

———, 1981, *Explosion Preventing Systems, National Fire Code*, Section 69, Volume 5, National Fire Protection Association, Boston, MA.

———, 1981, *Explosion Venting, National Fire Code*, Section 68, Volume 14, National Fire Protection Association, Boston, MA.

———, 1981, *Lightning Protection Code, National Fire Code*, Section 789, Volume 7, National Fire Protection Association, Boston, MA.

———, 1981, *Recommended Practice on Static Electricity, National Fire Code*, Section 77, Volume 15, National Fire Protection Association, Boston, MA.

________, 1981, General Indoor Storage, National Fire Code, Section 231, Volume 10, National Fire Protection Association, Boston, MA.

________, 1981, General Outdoor Storage, National Fire Code, Section 231A, Volume 15, National Fire Protection Association, Boston, MA.

Peckner, E., and I., Bernstein, 1977, Handbook of Stainless Steel, McGraw-Hill Publications, New York, NY.

Perry, R., and C. Chilton, 1973, Chemical Engineer's Handbook, McGraw-Hill Publications, New York, NY.

Petroleum Association for Conservation of the Canadian Environment (PACE), 1980, Buck Point Guidelines for Oil Spill Prevention and Control, Ottawa, Ontario.

Pludek, V.R., 1977, Design and Corrosion Control, John Wiley and Sons, New York, NY.

Rausch, A.H., N.A. Eisenberg, and C.J. Lynch, 1977, Continuing Development of the Vulnerability Model, Report No. CG-D-53-77, United States Coast Guard Office of Research and Development, Washington, DC.

Staniar, William, 1959, Plant Engineering Handbook, McGraw-Hill Publications, New York, NY.

Texas Air Control Board, 1981, Control of Air Pollution from Volatile Organic Compounds - Regulation V (31 TAC Chapter 115).

Thielsch, H., 1966, Defects and Failures in Pressure Vessels and Piping, Van Nostrand Rheinhold Company, New York, NY.

TI-3 Petroleum Committee of the Air Pollution Control Association, 1971, Informative Report No. 2, "Control of Atmospheric Emissions from Petroleum Storage Tanks," Journal of the Air Pollution Control Association, 21(5): 260-268.

Turner, D.B., 1970, Workbook of Atmospheric Dispersion Estimates, United States Public Health Service Publication Number 999-AP-26, Cincinnati, OH.

Underwriters' Laboratories, 1965, Storage of Flammable Liquids, U.L. File MH7991.

________, 1972, Steel Aboveground Tanks for Flammable and Combustible Liquids, UL 142, Northbrook, IL.

________, 1976, Steel Underground Tanks for Flammable and Combustible Liquids, UL 58, Northbrook, IL.

United States of America Standards Institute, 1969, Unfired Pressure Vessel Flange Dimensions, USAS B16.30, American National Standards Institute, New York, NY.

United States Department of Housing and Urban Development, 1975, Safety Considerations in Siting Housing Projects, Washington, D.C.

United States Department of Transportation, 1978, Emergency Action Guide for Selected Hazardous Materials, National Highway Traffic Safety Administration, Washington, DC.

United States Environmental Protection Agency, 1977, AP-42: Compilation of Air Pollution Emission Factors, Washington, D.C.

________, 1978, Control of Volatile Organic Emissions from Petroleum Liquid Storage in External Floating Roof Tanks, Washington, D.C.

________, 1977, Control of Volatile Organic Emissions from Storage of Petroleum Liquids in Fixed Roof Tanks, Washington, D.C.

________, 1981, Control of Volatile Organic Emissions from Volatile Organic Liquid Storage in Floating and Fixed-reef Tanks, Washington, D.C.

4. Corrosion Control

Corrosion is the process by which materials deteriorate to a more stable, natural state. Corrosion more than any other cause results in tank failures. Therefore, hazardous materials containment assurance programs should include adequate corrosion prevention measures.

An example of corrosion failure is depicted in Figure 4-1. In this situation, a tank inlet nozzle installed too close to the tank sidewall caused sulfuric acid to be sprayed on the tank wall at high velocity, which caused corrosion and erosion of the tank wall, and eventually the failure of the wall.

The position of inlet nozzles is only one isolated consideration which pertains to corrosion control. Other factors which must be considered include the compatibility of the storage tank with the materials being stored (see Section 2.3); environmental factors which may contribute to corrosion; selection of the most appropriate corrosion reduction methods; and use of tank maintenance and inspection methods which will not contribute to corrosion. Corrosion prevention requirements are determined by such factors as the desired lifetime of the tank, the conditions anticipated, and the desired purity of the product in the tank.

4.1 TYPES OF CORROSION

Corrosion appears in many forms. Recognizing which types of corrosion may occur in a given situation is an important step in corrosion prevention. The rest of this section lists and briefly describes the most common types of corrosion. Table 4-1 gives the percent frequency of occurrence of these types.

Uniform corrosion is a form of corrosion resulting in uniform deterioration across a metal surface. The uniform oxidation of a surface, such as rusting, is a common example of uniform corrosion.

Intergranular corrosion occurs when corrosion-promoting impurities have precipitated between the metal grains or crystals during fabrication. More serious than loss of metal is the general loss of strength and ductility associated with intergranular corrosion. Die-cast zinc aluminum alloys are particularly susceptible.

Pitting corrosion is localized corrosion which appears as pits or depressions in a metal surface. It commonly occurs when aluminium or

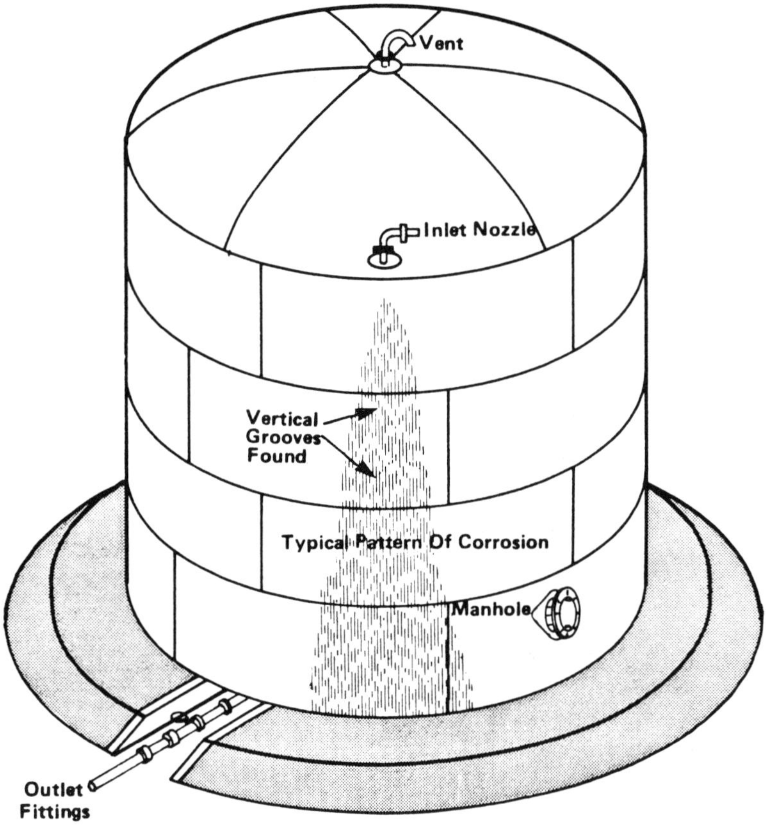

SOURCE: Shields and Dessert, 1981.

Figure 4–1 CORROSION DUE TO IMPROPER INLET NOZZLE PLACEMENT

Table 4-1

METAL FAILURE FREQUENCY OVER A TWO-YEAR PERIOD
(56.9% Corrosion and 43.1% Mechanical)*

Corrosion Failures†	Percent (%)
Uniform corrosion	31.5
Stress corrosion cracking, Corrosion fatigue	23.4
Pitting corrosion	15.7
Intergranular corrosion	10.2
Corrosion-erosion, Cavitation damage, Fretting corrosion	9.0
High temperature corrosion	2.3
Weld corrosion	2.3
Thermogalvanic corrosion	2.3
Crevice corrosion	1.8
Selective attack	1.1
Hydrogen damage	0.5

*The percentages can vary considerably in other industrial locations or environments.

Source: Pludek, 1977.

stainless steel surfaces are exposed to aqueous solutions containing chlorides. If uncorrected, pitting corrosion can lead to stress-corrosion cracking or corrosion fatigue.

Stress-corrosion cracking (Figure 4-2) is a form of corrosion in which external or internal residual stresses acting continuously in a corrosive environment cause cracks in the metal. Internal stress can result during metal cooling and from cold work such as pounding or bending. Aluminum alloys containing copper or magnesium and stainless steel, especially in high temperature steam service where steel pipes are wrapped in insulation or are in contact with chloride-containing solutions, are particularly susceptible to stress-corrosion cracking. Table 4-2 lists environments which cause stress-corrosion in various metals.

Corrosion fatigue arises when a metal is subjected to periodic stresses. The stress resistance of the material is gradually reduced as corrosion progresses. Corrosion fatigue occurs when metal oxide films which occur naturally on the surface of the metal are ruptured at a greater rate than new oxide films are formed. The rupture points become anodic, corroding preferentially and forming pits which can eventually lead to cracks in and ultimate failure of the metal.

Galvanic corrosion (Figure 4-3) occurs when dissimilar metals are in electrochemical contact through an electrolyte. Galvanic corrosion most commonly occurs when metal appurtenances are used on metal tanks. Table 4-3 shows an arrangement of common metals in the galvanic series. Anodic metals as a general rule are more easily corroded than noble or cathodic metals. Also as a general rule, the further apart two metals are in the series, the worse the threat of galvanic corrosion. The severity of galvanic corrosion is also determined by area relationships. The worst case occurs when a large area of a cathodic or noble metal is in contact with a small area of an anodic metal. For example, corrosion would be more likely if steel bolts were used on a copper tank than if copper bolts were used on a steel tank.

Thermogalvanic corrosion is a form of galvanic corrosion caused by a thermal gradient in a single piece of metal. Non-uniform heating causes different parts of the metal to have different electrochemical potentials, thus creating a galvanic cell conducive to corrosion.

Crevice corrosion is a problem because the intensity of corrosion is usually more severe in crevices between adjacent surfaces of either the same or different materials (one of them metal) than on surfaces. The increased corrosion rate is the result of a corrosion cell created by oxygen deficiency, acidity changes, ion buildup, or inhibitor depletion occurring in the crevice.

Oxygen-concentration cells (Figure 4-4) are localized areas of corrosion which are caused by differences in oxygen concentration. Areas of lower oxygen concentration, such as under gaskets or under solid residues on metal surfaces, are attacked. Oxygen-concentration cell corrosion can also be a form of crevice corrosion.

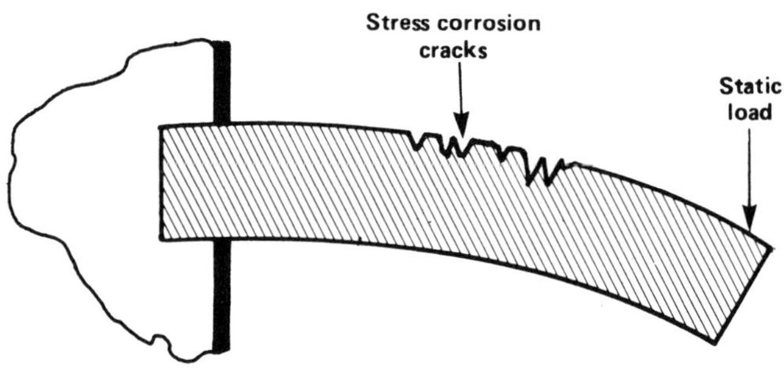

SOURCE: Pludek, 1977.

Figure 4–2 STRESS CORROSION CRACKING

Table 4-2

ENVIRONMENTS CAUSING STRESS CORROSION

Material	Environment
Aluminium	Water and steam; NaCl, including sea atmospheres and waters; air; water vapour
Copper	Tropical atmospheres; mercury; $HgNO_3$; bromides; ammonia; ammoniated organics
Aluminium bronzes	Water and steam; H_2SO_4; caustics
Austenitic stainless steels	Chlorides, including $FeCl_2$, $FeCl_3$, NaCl; sea environments; H_2SO_4; fluorides; condensing steam from chloride waters
Ferritic stainless steels	Chlorides, including NaCl; fluorides; bromides; iodides; caustics; nitrates; water; steam
Carbon and low alloy steels	HCl; caustics; nitrates; HNO_3; HCN; molten zinc and Na-Pb alloys; H_2S; H_2SO_4-HNO_3; H_2SO_4; seawater
High strength alloy steels (yield strength 200 psi plus)	Sea and industrial environments
Magnesium	NaCl, including sea environmenta; water and steam; caustics; N_2O_4; rural and coastal atmosphere; distilled water
Lead	Lead acetate solutions
Nickel	Bromides; caustics; H_2SO_4
Monel	Fused caustic soda; hydrochloric and hydrofluoric acids
Inconel	Caustics soda solutions; high purity water with few ppm oxygen
Titanium	Sea environments; NaCl in environments 288°C (550°F); mercury; molten cadmium; silver and AgCl; methanols with halides; fuming red HNO_3; N_2O_4; chlorinated or fluorinated hydrocarbons

Source: Pludek, 1977.

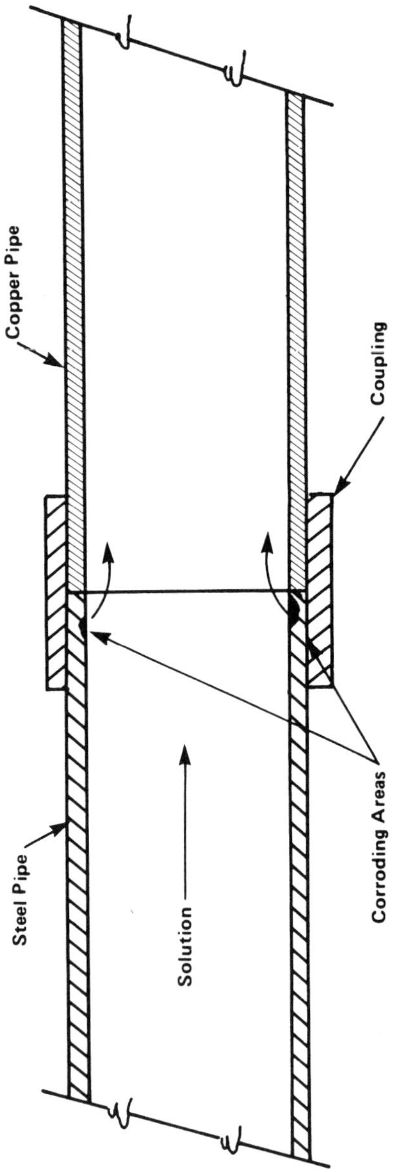

SOURCE: Department of the Navy, 1964.

Figure 4—3 INTERIOR GALVANIC CORROSION DUE TO COUPLING COPPER AND STEEL PIPE

Table 4-3

GALVANIC SERIES OF METALS AND ALLOYS

Corroded End (Anodic, or Least Noble)
Magnesium
Magnesium alloys
Zinc
Galvanized steel or galvanized wrought iron
Aluminum 6053
Aluminum 3003
Alulminum 2024
Aluminum
Alclad
Cadmium
Mild steel
Wrought Iron
Cast iron
Ni-Resist
13% chromium stainless (active)
50-50 lead-tin solder
18-6 stainless type 304 (active)
18-8-3 stainless type 316 (active)
Lead
Tin
Muntz metal
Naval brass
Nickel (active)
Inconel (active)
Yellow brass
Admiralty brass
Aluminum bronze
Red brass
Copper
Silicon bronze
70-30 cupronickel
Nickel (passive)
Inconel (passive)
Monel
18-8 stainless type 304 (passive)
18-8-3 stainless type 316 (passive)
Silver
Graphite
Gold
Platinum
Protected end (Cathodic, or Most Noble)

Source: Perry and Chilton, 1973.

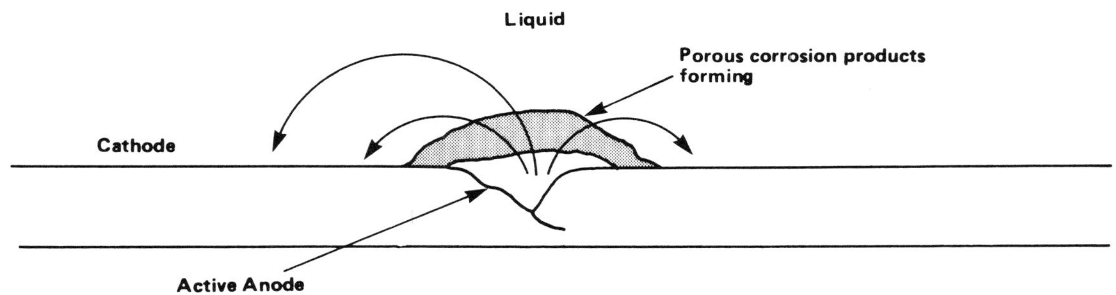

SOURCE: Department of the Navy, 1964.

Figure 4—4 OXYGEN CONCENTRATION CELL WITH RUST ON TANK WALL

Erosion is caused by the combination of corrosion and abrasion or friction when moving fluids are in contact with a metal surface. Abrasion occurs when the fluid contains suspended solids. In general, deterioration of metal is more severe with moving fluids than with stationary fluids.

Cavitation erosion is accelerated erosion occurring in agitated fluids when vacuum bubbles caused by the agitation collapse on the metal surface. The mechanical force of the collapsing bubbles can destroy protective oxide films and result in severe localized pitting before new oxide films can form.

Impingement attack (Figure 4-5) is localized corrosion/erosion due to turbulent flow of liquids. It is especially severe at conduit entrances and bends in pipes.

Mechanical corrosion is deterioration caused by repeated mechanical impingement on a metal surface. A common example is corrosion of tank bottoms at points where measuring sticks used for level gauging repeatedly strike the inside metal surface. This type of mechanical corrosion is remedied by installation of strike plates at the bottom of the tank so that uniform corrosion patterns will not be distrubed.

Fretting corrosion is caused by friction between metal surfaces. It usually occurs under heavy loads and only slight metal movement, such as that induced by high frequency vibrations. Frictional heat causes oxidation of the metal, and the metal oxides which are formed are removed by the friction itself. This exposes a fresh metal surface, which in turn is oxidized and rubbed off.

Hydrogen embrittlement refers to the general decrease in strength and ductility of a metal which occurs when hydrogen is absorbed into the metal. In some cases, the hydrogen can react with impurities in the metal, such as carbon in carbon steel. Sources of hydrogen are cleaning processes, pickling and other treatment processes, and welding.

Stray current corrosion occurs when a direct current deviates from its intended path and passes through a conducting metal by way of an electrolyte. Corrosion generally occurs at the point where the current leaves the metal. Common sources of stray current are underground cables, or DC machinery.

Differential environment cells (Figure 4-6) are corrosion cells which are established at the interface between two different media in contact with the same piece of metal. Corrosion of this type can occur at air-water interfaces or at interfaces between different soil types.

4.2 ENVIRONMENTAL FACTORS AFFECTING CORROSION

The characteristics of the soil, water, or air surrounding a tank or pipeline have an important effect on the type and degree of corrosion which will occur in a given situation. While often there is no

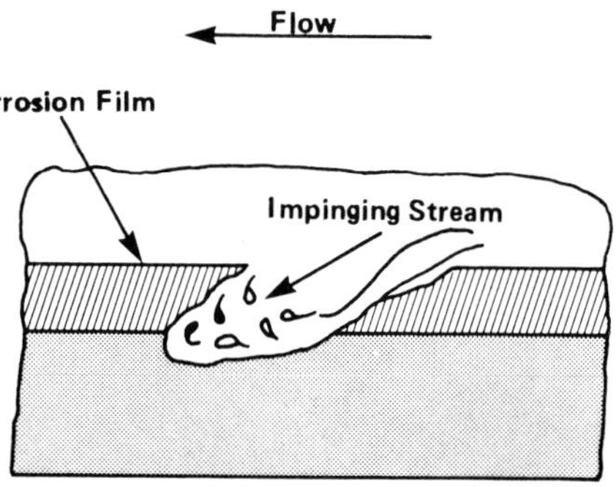

SOURCE: Pludek, 1977.

Figure 4–5 IMPINGEMENT ATTACK

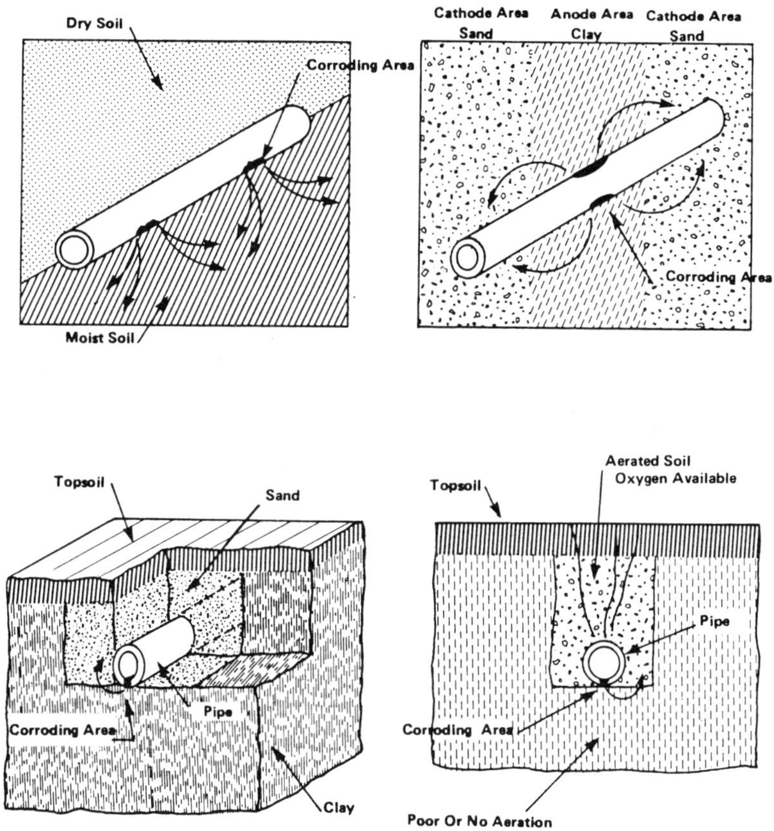

SOURCE: Department of the Navy, 1964.

Figure 4–6 DIFFERENTIAL ENVIRONMENT UNDERGROUND

practical way to alter these conditions, recognizing the general characteristics of the surrounding environment provides a basis for selecting an appropriate corrosion control system.

Soil characteristics which promote corrosion include non-uniform composition, poor aeration (e.g., as in clay), high or low pH, high organic content, anaerobic bacteria, low or non-uniform resistivity, and high moisture content.

Aqueous environments are generally more corrosive with high temperatures, high dissolved oxygen content, agitation, highly alkaline or acid pollution, and low resistivity.

The corrosivity of atmospheric environments is typically increased with increased humidity, increased temperature, and pollution.

4.3 CORROSION CONTROL METHODS

Corrosion control can range from the common sense practice of avoiding obviously corrosive conditions, such as galvanic couples or chemical/material incompatability, to recognized control systems such as cathodic protection. The most common corrosion control systems are corrosion inhibitors in stored liquids, cathodic or anodic protection, linings, and coatings. As the terms are used here, linings refer to protection to tank and pipe interior surfaces, while coatings refer to exterior surface protection. In general, thin (less than 10 mils) brush- or spray-applied protection is adequate for tank and pipe exteriors, but not for interiors, which are exposed to the full effect of corrosive materials. Linings generally must be on the order of 100 mils or greater. Two or more control methods are often combined to insure maximum protection. Table 4-4 lists the common forms of corrosion and general methods for controlling them.

Before specific corrosion control methods are applied, a general understanding of corrosion potential under certain circumstances is essential. Taking this potential into account during equipment and facility design and installation can significantly reduce incidences of corrosion.

4.3.1 Protective Liners

A chemically resistant tank can often be achieved economically by the use of a protective lining in a relatively inexpensive base metal tank. Ferrous metal tanks are commonly used in this application. The selection of the lining material depends mainly on the chemical to be stored, the storage temperature, and the extent and type of abrasion to which the lining will be exposed. If possible, field performance and ease of application should also be reviewed before a particular lining material is chosen. Using Appendix C, chemical compatibility, the prime consideration in lining selection, can be checked for many common lining materials.

Common lining materials are epoxy resins, furane resins, phenolic resins, polyethylene, PVC, saran, rubber, glass, ceramic, and

Table 4-4

CORROSION CONTROL METHODS

Type of Corrosion	Control Methods
Uniform Corrosion	• Inhibitors • Protective coating • Anodic protection
Intergranular Corrosion	• Avoiding temperatures that can cause contaminant precipitation during heat treatment or welding
Pitting Corrosion	• Protective coating • Allowing for corrosion in wall thickness
Stress-Corrosion Cracking	• Reducing residual or applied stresses • Redistributing stresses • Avoiding misalignment of sections joined by bolts, rivets, or welds • Materials of similar expansion coefficients in one structure • Protective coating • Cathodic protection
Corrosion Fatigue	• Minimizing cyclic stresses and vibrations • Reinforcing critical areas • Redistributing stresses • Avoiding rapid changes in load, temperature, or pressure • Inducing compressive stresses through peening, swagging, rolling, vapor blasting, chain tumbling, etc.
Galvanic Corrosion	• Avoiding galvanic couples • Completely insulating dissimilar metals (Paint alone is insufficient) • Using filler rods of same chemical composition as metal surface during welding • Avoiding unfavorable area relationships • Using replaceable parts of the anodic (attacked) metal • Cathodic protection • Inhibitors

Table 4-4 (Cont.)

Type of Corrosion	Control Methods
Thermogalvanic Corrosion	• Avoiding non-uniform heating and cooling • Maintaining uniform coating or insulation thickness
Crevice Corrosion; Concentration Cells	• Minimizing sharp corners and other stagnant areas • Minimizing crevices to a minimum, especially in heat transfer areas and in aqueous environments containing inorganic solutions or dissolved oxygen • Enveloping or sealing crevices • Protective coating • Removing dirt and mill-scale during cleaning and surface preparation • Welded butt joints with continuous welds instead of bolts or rivets • Inhibitors
Erosion; Impingement Attack	• Decreasing fluid stream velocity to approach laminar flow • Minimizing abrupt changes in flow direction • Streamlining flow where possible • Installing replaceable impingement plates at critical points in flowlines • Filters and steam traps to remove suspended solids and water vapor • Protective coating • Cathodic protection
Cavitation Damage	• Maintaining pressure above liquid vapor pressure • Minimizing hydrodynamic pressure differences • Protective coating • Cathodic protection • Injecting or generating larger bubbles
Fretting Corrosion	• Installing barriers which allow for slip between metals • Increasing load to stop motion, but not above load capacity • Porous protective coating • Lubricant

Table 4-4 (Cont.)

Type of Corrosion	Control Methods
Hydrogen Embrittlement	• Low-hydrogen welding electrodes • Avoiding incorrect pickling, surface preparation, and treatment methods • Inducing compressive stresses • Baking metal at 200-300°F to remove hydrogen • Impervious coating such as rubber or plastic
Stray-current Corrosion	• Providing good insulation on electrical cables and components • Grounding exposed components of electrical equipment • Draining off stray currents with another conducting material • Electrically bonding metallic structures • Cathodic protection
Differential-environment Cells	• Underlaying and backfill underground pipelines and tanks with the same material • Avoiding partially buried structures • Protective coating • Cathodic protection

Source: Adapted from Pludek, 1977.

concrete. Inorganic zinc coatings are applied mainly in the storage of organic solvents, aromatics, ketones, hydrocarbons, and water.

The use of a lining material imposes certain restrictions on tank construction. In general, lined metal tanks should be of welded construction, and any buttwelds, welding clusters, or other metal protrusions should be ground smooth prior to lining. Filler welds should be used on corners, and all angles should have radii of at least 1/4 inch. Surface preparation is also very important. Lined surfaces should be white metal blasted in accordance with Steel Structure Painting Council (SSPC) Specification SP-5, or National Association of Corrosion Engineers Specification 1. The prepared surface should ultimately be free of weld spaths, pits, grease, dirt, and rust, and should not be painted or galvanized before being lined. The lining material should be applied as soon as possible after blasting.

Linings can be applied in several ways. When applied in sheets, as are most rubber linings, an appropriate adhesive is brushed or sprayed on the surface, and the lining laid down and rolled on by hand. It is essential when applying sheets that no air pockets remain under the lining. Air pockets are most likely to remain at corners and seams, and thus it is common to press the lining at these points with hot iron tools to make thorough contact. Linings can also be applied by dipping the metal into a bath of lining material and re-dipping until the desired thickness is attained. Dipping is useful for lining odd-shaped structures, but the usefulness is limited by the difficulty of obtaining a uniform thickness of the liner material. Glass, plastic, and ceramics are sometimes applied in powder form and then heated to fuse the lining material to the metal. Concrete linings are applied by centrifugal casting, trowelling, and spraying. Thick concrete linings are generally reinforced with wire mesh.

Some lining materials such as glass, ceramic, and concrete are of limited usefulness because of their sensitivity to mechanical damage and thermal shock. Thus, handling of lined equipment is the most common source of damage to the linings. Another disadvantage of concrete linings is that the wire mesh used as reinforcement is also subject to corrosion. Once they are in place, lined tanks should always be spark-tested, as discussed in Section 5.2.4

4.3.2 Protective Coatings

Corrosion of tank exteriors can be controlled by the use of resistant construction materials and paints. The required degree of control depends mainly on atmospheric and use conditions to which the tank is exposed. Most paints, for instance, do not provide adequate protection of underground tanks and pipelines that are not otherwise protected. The SSPC specification designates four classes of Standard Corrosive Environments, ranging from highly humid, industrial conditions with harsh chemical and weather exposure, to dry rural conditions with no chemical exposure. Table 4-5 lists the most common paints and their relative resistances to major chemical groups, weather, abrasion, and heat. This table should be used only for general information on the various coatings. More detailed information should be obtained from paint manufacturers.

Table 4-5

COMPARATIVE RESISTANCES OF TYPICAL COATINGS

Coating Type	Acid	Alkali	Salts	Solvents	Water	Uxidation	Sunlight and Water	Stress	Abrasion	Heat
Acrylic	8	8	9	5	8	9	10	?	10	8
Alkyd	6	6	8	4	8	3	10	5	6	8
Asphalt	10	7	10	2	10	2	7	5	3	4
Chlorinated Rubber	10	10	10	3	10	9	7	7	7	5
Epoxy	10	9	10	8	10	6	9	3	6	9
Furan	10	10	10	10	10	2	8	1	5	9
Inorganic (metallic)	1	1	5	10	5	10	10	?	10	10
Latex	2	1	6	1	2	1	10	?	6	5
Neoprene	10	10	10	4	10	6	8	10	10	10
Oil Base	1	1	6	2	7	1	10	4	4	7
Phenolic	10	2	10	10	10	7	9	2	5	10
Saran	10	8	10	5	10	10	7	7	7	7
Urethanes	9	10	10	9	10	9	8	?	10	8
Vinyl	10	10	10	5	10	10	10	8	7	7

Scale: 1 = Nonresistant
10 = Extremely resistant
? = Insufficient data

Sources: NACE, 1975, and Staniar, 1959.

The most important step in exterior surface coating is surface preparation. The three major types of surface preparation, in increasing order of achievable surface quality, are hand-tool cleaning, power-tool cleaning, and sand blasting. In general, surface preparation requirements for maintenance painting are not as stringent as those for initial painting, unless complete repainting is being performed. The SSPC specifications can be consulted for detailed guidelines.

In some cases, especially with steel, the surface to be painted is given a chemical conversion coating as a primer. The chemical conversion coating is a solution, such as a phosphate, which forms crystals that bond to the metal. Such coatings should not be used without a further coating, however.

The two main methods of applying protective coatings are brushing and spraying. Most coatings are designed for spray application, because spraying is generally neater, more uniform, and faster than brushing. However, first coats are often brush-applied because better surface-wetting can be achieved by brushing, and because air pockets can form if first coats are spray-applied. The most important requirements for the application of coating are that the coat be evenly applied and that it be applied as soon as possible after surface preparation. For instance, painting schedules should be arranged so that no prepared surface will go uncoated overnight.

The initial maintenance of a new paint system should occur within the first six months, so that inadequacies can be rectified before serious damage takes place. Subsequent maintenance painting should follow a planned program, so that major painting jobs do not all occur at the same time. Five-year plans are commonly used for maintenance painting, although the specific program for a facility will depend on its size, the paint system used, the environmental conditions, housekeeping practices, and other considerations.

4.3.3 Cathodic Protection

Cathodic protection is used to eliminate corrosion cells on a metal surface by means of an externally applied current which opposes the corrosion potential of the protected metal. It has wide application for aboveground and underground tanks and pipelines containing a wide range of chemicals, and it is particularly important in that it is usually the only practicable means for halting corrosion already in progress. The two established means of providing cathodic protection are the sacrificial anode method and the impressed current method.

Sacrificial anodes are magnesium, zinc, or aluminum electrodes which are electrically connected to the protected metal in an electrolytic environment. Typical installations are shown in Figure 4-7. Systems employing sacrificial anodes are designed so that the sacrificial anode is more anodic than the entire protected structure and will thus corrode preferentially. The main limitation of this method is that the current which is established through the presence of the sacrificed anodes is sometimes not strong enough to provide adequate

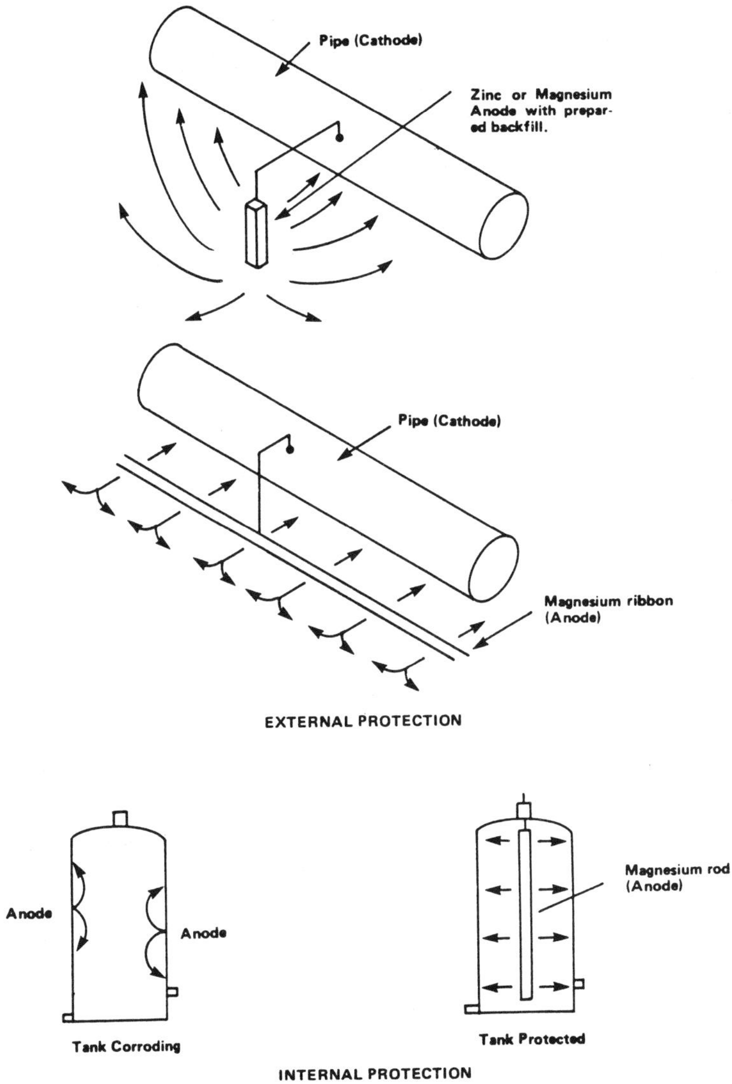

SOURCE: Department of the Navy, 1964.

Figure 4–7 CATHODIC PROTECTION BY THE SACRIFICIAL ANODE METHOD

protection, especially in highly resistive electrolytic media. For this reason, sacrificial anodes are most often used in conjunction with relatively small installations, for localized protection of "hot spots" on a structure, and with well-coated structures which do not require as great a current for protection as bare or poorly coated metal. Sacrificial anodes are also preferred in locations where the electrical equipment required for the impressed-current method poses a hazard.

The impressed current method requires an external source of direct current which is transmitted to the electrolyte (i.e., soil or liquid tank contents) through anodes made of graphite, carbon, scrap iron or steel, aluminum, platinum, or silicon cast iron. An example of an impressed-current arrangement is given in Figure 4-8. Because one installation of an impressed-current system can protect a large metal surface area, this method is typically used for large systems, for bare or poorly coated surfaces, or when the electrolyte is highly resistive. The main drawbacks are the potential for stray currents, which can induce electrolytic corrosion on structures near the one being protected, and overprotection, which can lead to hydrogen attack of the protected metal as hydrogen accumulates on the surface.

The current output requirement of the anodes in either method is a function of the current density required at the metal surface, the metal surface area, the resistivity of the electrolyte, the size and type of the anodes, the spacing of the anodes, and the distance between the anode and the surface. Current requirements become especially complicated in close-packed storage systems in which current interference can hamper cathodic protection. The determination of these factors requires a certain amount of testing and calculations before a protection system is selected. Procedures can be found in NAVDOCKS MO-306 and in many other references dealing with corrosion.

The use of cathodic protection requires adequate electrical bonding or insulation of protected structures. Bonding of adjacent sections of a protected structure through the use of low-resistance straps or cables is necessary to insure that the impressed current is applied to all areas requiring protection. Conversely, the protected structure should be electrically isolated from adjacent structures and appurtenances through the use of insulated joints, flanges, and dielectric bushings. Electrical isolation prevents the impressed current from overreaching its intended range.

Maintenance requirements for cathodic protection systems depend on the specific type of protection. Sacrificial anodes generally are designed to last up to 20 years, while impressed current systems should be checked every four to six years. The electrical equipment associated with the impressed-current method requires routine maintenance throughout the lifetime of the system.

4.3.4 Anodic Protection

Anodic protection is similar to cathodic protection, but with anodic protection the protected metal (usually iron or steel) is made

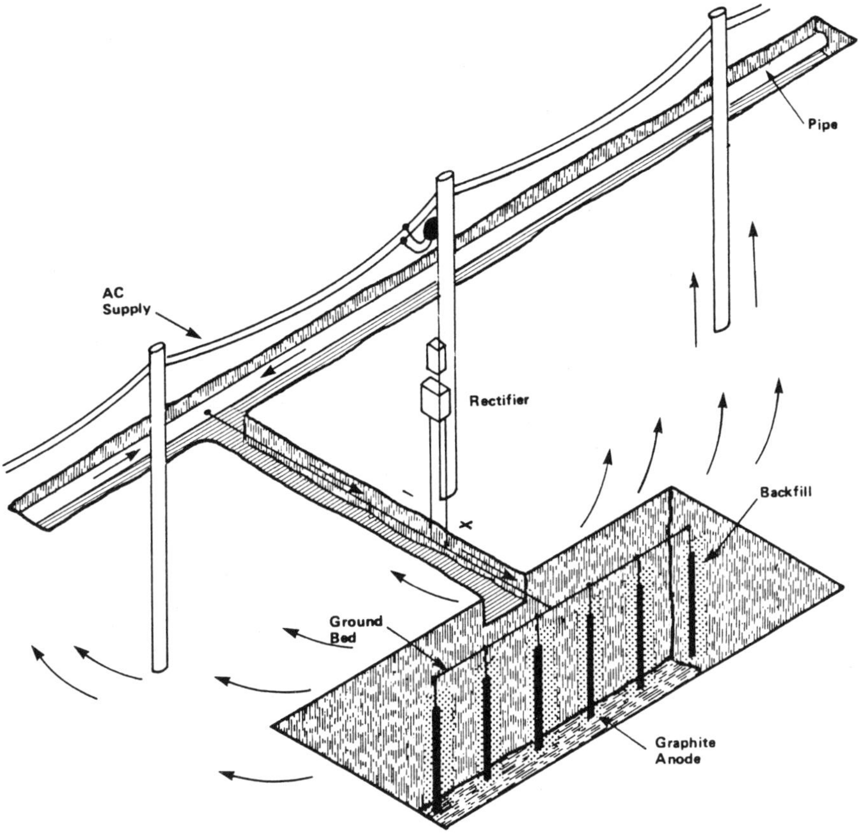

SOURCE: Department of the Navy, 1964.

Figure 4–8 CATHODIC PROTECTION BY THE IMPRESSED CURRENT METHOD

anodic instead of cathodic. The impressed current in this case increases the passivity of the metal beyond the level associated with normal oxide film formation.

The most common application of anodic protection is in sulfuric acid storage in steel tanks, although it is also applicable with phosphoric and other acids, some alkali solutions, and some salt solutions, such as sodium sulfate and sodium nitrate. In general, however, anodic protection is much less common than cathodic protection.

Anodic protection of steel and iron is limited by the presence of chloride or other halide ions, although anodic protection can be used to protect titanium against corrosion by hydrochloric acid. Anodic protection does not work with metals such as zinc, magnesium, copper, and copper alloys, because they do not become sufficiently passive when anodically polarized.

4.3.5 Inhibitors

Corrosion on the inside of a storage tank can sometimes be controlled by adding a corrosion inhibitor to the stored liquid. Typical inhibitors are chromates, phosphates, silicates, organic sulfides, and amines. Selection of the appropriate inhibitor in a particular situation is dependent on the material being protected, the stored liquid, and the storage conditions. Table 4-6 lists some common inhibitors and their usual applications. It is essential that the correct concentration of the inhibitor be used. Concentrations that are too small can increase corrosion, resulting in extensive pitting in severe cases. The correct minimum concentration is difficult to determine because it is dependent on surface quality of the tank or appurtenances, the nature of the corrosive liquid, and other factors. Too much inhibitor is also undesirable. Once the correct concentration is established, it should be checked continually, because the inhibitor can disappear through absorption, chemical degradation, decomposition, precipitation, or evaporation. The use of inhibitors generally necessitates agitation of the liquid, and this can be an additional limiting factor.

4.3.6 Compressive Stress Induction

Compressive stress can be induced to combat corrosion fatigue and stress corrosion cracking. Susceptible materials such as steel and aluminum can be strengthened by treating the surface by shot peening, chain tumbling, vapor blasting, or an equivalent method.

4.3.7 Strikeplates

Strikeplates are sheets of metal placed inside a tank to prevent mechanical corrosion of tank bottoms due to repeated mechanical impingement, such as of measuring sticks on metal surfaces. Strikeplates must be placed carefully to prevent other forms of corrosion, such as those discussed in Section 4-1.

Table 4-6

TYPICAL INHIBITORS AND THE CORROSION ENVIRONMENT IN WHICH THEY ARE EFFECTIVE

Inhibitor	Inhibitor Concentration (percent by weight)	Environment	Metals to be Protected
Benzanilide	0.2	Lubricants	Cd-Ni: Co-Pbb bearings
Borax	2-3	Alcohol anti-freeze	Car cooling systems
Calgon	small	Water	Steel
Dioctyl ester of sulpho-succinic acid	0.05	Refined petroleum	Pipelines
Disodium hydrogen phosphate	0.5	Citric acid	Steel
Erythritol	small	K_2SO_4	Mild steel
Ethylaniline	0.5	HC solutions	Ferrous metals
Formaldehyde	small	Oil wells	Oil-well equipment
Mercaptobenzthiazole	1	HC solutions	Iron and steel
Morpholine	0.2	Water	Heat exchangers
Oleic acid	small	Polyhydric alcohol	Iron
Phenyl acridine	0.5	H_2SO_4 solutions	Iron
Potassium dichromate	0.05-0.2	Tap water	Iron-brass
Potassium dihydrogen Phosphate + sodium nitrate	small + 5 percent	Seawater	Steel
Potassium permanganate	0.1	NaOH solutions	Aluminum
Pyridine + phenylhydrazine	0.5 + 0.5	HC solutions	Ferrous metals
Quinoline ethiodide	0.1	H_2SO_4	Steel
Rosin amine-ethylene oxide	0.2	HC solutions	Mild steel
Sodium benzoate	0.5	NaCl solutions	Mild steel
Sodium carbonate	small	Condensate	Iron
Sodium chromate	0.07	$CaCl_2$ brine cooling water	Copper-brass rectifiers
Sodium dichromate	0.025	Water	Air conditioning
Sodium dichromate + sodium nitrate	0.1 + 0.05	Water	Heat exchangers

Table 4-6 (Cont.)

Inhibitor	Inhibitor Concentration (percent by weight)	Environment	Metals to be Protected
Sodium hexametaphosphate	0.002	Water (about pH 6)	Lead
Sodium metaphosphate	Small	Ammonia	Mild steel
Sodium nitrite	0.005	Water	Mild steel
Sodium orthophosphate	1	Water (pH 7.25)	Iron
Sodium silicate	Small	Seawater	Zn; Zn-Al alloys
Tetramethylammonium oxide	0.5	Aqueous solutions of organic solvents	Iron and steel
Thiourea	1	Acids	Iron and steel

Source: Uhlig, 1971.

4.4 ESTABLISHING A CORROSION CONTROL PROGRAM

The implementation of the anti-corrosion measures mentioned in this section should be incorporated in a corrosion control program. Such a program should include:

- Identification of the material of construction of all storage tanks and other equipment involved;
- Identification of construction methods used for equipment involved;
- Determination of potential corrosion-related failures and associated hazards;
- Compilation of historical data on corrosion-related failures;
- Compilation of atmospheric data which could affect corrosion;
- Compilation of soil data related to underground corrosion;
- Determination of appropriate preventive measures for each piece of equipment;
- Identification of experts for consultation on corrosion problems; and
- Keeping comprehensive maintenance records.

BIBLIOGRAPHY

American Institute of Chemical Engineers, *et al.*, 1975, *Managing Corrosion Problems with Plastics*, National Association of Corrosion Engineers, Houston, TX.

Department of the Navy, 1964, *Corrosion Prevention and Control*, NAVDOCKS MO-306, Washington, D.C.

Edeleanu, C., 1978, Safety in Process Plant in Relation to Corrosion, *Chemistry and Industry*, 17:649-651.

Evaus, V., 1961, *The Corrosion and Oxidation of Metals*, St. Martins Press, New York, NY.

__________, 1968, *The Corrosion and Oxidation of Metals: First Supplement*, St. Martins Press, New York, NY.

Fitzgerald, J.H., and A.L. Claes, 1976, *Fundamentals of Underground Corrosion Control*, Technical Publishing, New York, NY.

Hepner, I., ed., 1962, *Materials of Construction for Chemical Plant*, Leonard Hill Ltd., London.

McDonnell, C., 1979, Erosion/Corrosion of Plant Machinery in Fluid Flow Conditions, *Anti-Corrosion Methods and Materials*, 26(10): 5-6, 15.

Mellan, I., 1976, *Corrosion Resistant Materials Handbook*, Noyes Data Corporation, Park Ridge, NJ.

Menke, J., 1981, Ten Commandments of Material Deterioration, *Materials Performance*, 120(4):21-23.

Morris, D., 1980, Plastic Linings - Specifying for Performance, *Corrosion Prevention and Control*, 27(6):19-20.

NACE, 1954, *Corrosion Data Survey for Metals*, National Association of Corrosion Engineers, Houston, TX.

__________, 1975, *Process Industries Corrosion*, National Association of Corrosion Engineers, Houston, TX.

__________, 1976, *Corrosion Data Survey for Non-Metals*, National Association of Corrosion Engineers, Houston, TX.

__________, 1980, Corrosion Engineer's Reference Book, R.S. Teseder, ed., National Association of Corrosion Engineers, Houston, TX.

Perry, R., and C. Chilton, 1973, Chemical Engineer's Handbook, McGraw-Hill Publications, New York, NY.

Pludek, V.R., 1977, Design and Corrosion Control, John Wiley and Sons, New York, NY.

Rabald, E., 1968, Corrosion Guide, Elsevier Scientific Publishing Company, New York, NY.

Schields, E., and W.J. Dessert, 1981, Learning A Lesson From a Sulfuric Acid Tank Failure, Pollution Engineering, 13(12):39-40.

Shreir, L., 1976, Corrosion, Volumes I and II, Newnes-Butterworth Publishing Company, Woburn, MA.

Staniar, William, 1959, Plant Engineering Handbook, McGraw-Hill Publications, New York, NY.

Whlig, H.H., 1971, Corrosion and Corrosion Control, John Wiley and Sons, New York, NY.

Wilson, C.L., and J.A. Oates, 1968, Corrosion and the Maintenance Engineer, Hart Publishing Company.

5. Tank Inspection and Maintenance

The goals of an appropriate inspection and maintenance program for chemical storage tanks are:

- To minimize the probability of accidental releases of hazardous materials;
- To reduce the risks of fire and exposure resulting from such releases;
- To maintain safe working conditions in and around the storage area; and
- To permit early detection of potential trouble spots and implement corrective action.

To accomplish these goals an inspection program must be implemented which is able to identify excessive corrosion or erosion, structural fatigue or cracking of metals, deterioration of non-metallic liners and appurtenances, cracking or weakening of welds and joints, and leakage from valves or piping. Special attention should be paid to likely trouble spots, which include bottom-to-shell connections; valve, nozzle, and manhole connections; welded seams; rivet holes; and welded brackets.

5.1 NON-DESTRUCTIVE TESTING

Numerous methods of non-destructive testing are available to accomplish a variety of inspection goals, such as leak detection, wall thickness measurement, or checking liner integrity. Most of these methods are applicable to aboveground tanks and to underground tanks prior to installation. A discussion of some of the major methods and their limitations follows. Standards for these methods may be found in Section V of the ASME Boiler and Pressure Vessel Code. Table 5-1 lists the types of imperfections detectable by the various methods. Test methods for specific types of storage tanks and equipment are discussed in Section 5.2.

5.1.1 Test Methods

- Radiographic Testing - Radiographic testing employs the use of X-rays, nuclear radiation, or both to detect subsurface discontinuities in solid materials, and to present their images

Table 5-1

APPLICATION OF NON-DESTRUCTIVE TEST METHODS

Type of Imperfection Detected	Test Type									
	Visual	Random Radiographic	100% Radiography	Ultrasonic	Wet Magnetic Particle	Dry Magnetic Particle	Liquid Penetrant	Hydrostatic	Eddy Current	Spark Testing
VALVE										
Cracks	X				X					
Strength								X		
TANK										
Cracks or surface discontinuties	X				X	X	X		X	X
Subsurface dis-continuties		X		X					X	X
Thinning				X					X	
Strength								X		
WELDS										
Crack	X				X	X	X			
Incomplete penetration	X	X	X							
Porosity			X							
Slag inclusions			X							

SOURCE: Ecology and Environment, Inc., 1982.

on a recording medium (film), known as a radiograph. Any flaws detected by the test will appear as darkened areas, in the shape of the flaw, against the uniformly lighter background of the intact area.

- Ultrasonic Testing - Ultrasonic testing detects subsurface discontinuities from the interruptions they cause in pulse or resonant vibrations transmitted through the material. The ultrasonic waves are transmitted through the metal until they reach a reflecting surface, which returns the waves. The time interval required for the waves to complete this "round trip" indicates the metal thickness. This testing method can be used on metal in a range of thicknesses from a fraction of an inch to several feet.

- Wet Magnetic Particle Testing - Magnetic particle inspection is used to detect surface cracks or flaws. Fine magnetic particles are applied to a magnetized surface and are attracted to regions of magnetic nonuniformity associated with such cracks or discontinuities. Indicative patterns arise which can be observed visually. The wet method is adaptable to irregular, relatively small surface areas, and is thus ideal for valves.

- Dry Magnetic Particle Method - The dry particle method is similar in principle to the wet method. It is more sensitive than the wet method in the detection of near-surface discontinuities, but less sensitive in detecting fine surface discontinuities. The associated equipment is portable and well-suited to field work, and is more applicable to large surface areas, such as tank shells, than the wet method.

- Liquid (Dye) Penetrant Testing - The dye penetrant method involves applying to a surface a liquid which will seep into any surface cracks or discontinuities through capillary action. After the surface is wiped dry, a developer is applied which becomes tainted by the original liquid as it seeps out of the cracks, thus delineating the cracks or discontinuities. The method is used effectively on nonporous metallic materials, both ferrous and nonferrous, and on nonporous, nonmmetallic materials such as ceramics, plastics, and glass.

- Hydrostatic Testing - The hydrostatic test can reveal gross flaws, inadequate design, and flange leaks. It is most valuable as a test on new tanks before they are put into service, and certification of such should be requested of the manufacturer by purchasers of new tanks. A simple standpipe test is useful for determining gross tank leakage. The more sensitive Kent-Moore test compensates for changes in temperature, pressure, and viscosity, and can detect leaks as small as 0.05 gal/hr.

- Eddy Current Test - Eddy currents are electrical currents induced within the body of a conductor when that conductor moves through a nonuniform magnetic field or is in a region where there is a change in magnetic flux. In the eddy current test, a test coil is brought close to a conducting specimen. Changes of impedance of the test coil indicate the eddy currents induced by the coil, thereby indicating defects within the specimen. The method is effective for spot checks of surface and subsurface cracks, wall thickness, and coating thickness.

- Spark Testing - High voltage, low current electrical spark tests are performed by passing the electrode over a nonconducting material, such as a tank lining or coating. The other end of the circuit is attached to the conductive tank wall. Any "holidays" (defects) in the lining will cause an electrical arc to pass through at the point of the defect. Care must be taken not to exceed the dielectric, or damage to the lining may result.

5.1.2 Quality Control

Wherever possible, it is desirable to enhance quality control by optimizing the test procedure by checking against available references. For example, the radiographic test makes use of standard penetrameters to obtain evidence on a radiograph that the technique used was satisfactory. A desired level of test quality or sensitivity is established, and then tested by confirming that penetrameter openings of known size are reproducible on the radiograph. These penetrameters are not intended for use in estimating the size of discontinuities detected, but rather for quality optimization. Similarly, discontinuities detected in ferrous castings by the dry particle method may be compared to reference photographs contained in "ASTM Reference Photographs E125, for Magnetic Particle Indications on Ferrous Castings."

Detailed specifications and procedures to insure quality control for all nondestructive tests are contained in the referenced ASME standards.

Table 5-2 provides a listing of nondestructive tests including those discussed in Section 5.1.1. The table summarizes the types of defects or measurements determined by the tests, their applications, as well as their advantages and limitations. Although the table provides guidelines for determining the effectiveness of the tests, qualified operators and conformance with manufacturers' specifications and the appropriate ASME and ASTM standards are necessary to achieve optimal results.

5.2 INSPECTION PROCEDURES

5.2.1 Aboveground Tanks

The extent of inspection procedures for aboveground tanks depends on whether the tank is out of service or in service. Inspection of in-service tanks will necessarily be restricted to exterior surfaces and appurtenances. Gross leakage will be readily evident, but closer

Table 5-2

NON-DESTRUCTIVE TEST METHODS

Method	Measures or Detects:	Applications	Advantages	Limitations
Acoustic emission	Crack initiation and growth rate	Pressure vessels	Remote and continuous surveillance	Transducers must be placed in contact with surface of part to be tested
	Internal cracking in welds during cooling	Stressed structures	Permanent record	Highly ductile materials yield low amplitude emissions
	Boiling or cavitation		Dynamic (rather than static) detection of cracks	Part must be stressed or operating
	Friction or wear		Portable	Test system noise needs to be filtered out
	Plastic deformation		Triangulation techniques to locate flaws	
Acoustic-impact (tapping)	Debonded areas or delaminations in metal or non-metal composites or laminates	Brazed or adhesive-bonded structures	Portable	Part geometry and mass influences test results
	Loose rivets or fasteners	Bolted or riveted assemblies	Easy to operate	Impactor and probe must be repositioned to fit geometry of part
	Crushed core	Composite structures	May be automated	Reference standards required
		Honeycomb assemblies	Permanent record or positive meter readout	Pulser impact rate is critical for repeatability
			No couplant required	

Table 5-2 (Cont.)

Method	Measures or Detects:	Applications	Advantages	Limitations
Eddy current (200 Hz to 6 MHz)	Surface and subsurface cracks and seams	Tubing	No special operator skills required	Conductive materials
	Alloy content	"Spot checks" on all types of surfaces	High speed, low cost	Shallow depth of penetration (thin walls only)
	Heat treatment variations	Proximity gage	Automation possible for symmetrical parts	Masked or false indications caused by sensitivity to variations such as part geometry, lift-off
		Metal detector		
	Wall thickness, coating thickness	Metal sorting	Permanent record capability for symmetrical parts	Reference standards required
	Crack depth	Measure conductivity	No couplant or probe contact required	Permeability variations
	Conductivity			
	Permeability			
Electric current	Cracks	Metallic materials	Access to only one surface required	Edge effect
	Crack depth	Electrically conductive materials		Surface contamination
			Battery or DC source	
	Resistivity			Good surface contact required
			Portable	
	Wall thickness			Difficult to automate
	Corrosion-induced wall thinning			Electrode spacing
				Reference standards required

Table 5-2 (Cont.)

Method	Measures or Detects:	Applications	Advantages	Limitations
Fluoroscopy (Cine-fluorography) (Kine-fluorography)	Level of fill in containers Foreign objects Internal components Density variations Voids, thickness Spacing or position	Particles in liquid flow Presence of cavitation Operation of valves and switches	High-brightness images Real-time viewing image magnification Permanent record Moving subject can be observed	Costly equipment Lack of geometric sharpness Thick specimens Speed of event to be studied Viewing area
Holiday detector High voltage (spark)	Inegrity of coatings or linings	Detects holidays in coatings of thickness >15 mils	Portable Easy to operate	Possible damage if dielectric strength exceeded
Holiday detector Low voltage	Integrity of coatings	Detects holidays in coatings of thickness <20 mils	Portable Easy to operate	Requires contact with substrate
Leak testing	Leaks Helium, Ammonia, Smoke, Water, Air Bubbles, Radioactive gas, Halogens	Joints Welded, Brazed, Adhesive-bonded Sealed assemblies Pressure or vacuum chambers Fuel or gas tanks	High sensitivity to extremely small, tight separations not detectable by other NDT methods Sensitivity related to method selected	Accessibility to both surfaces of part required Smeared metal or contaminants may prevent detection Cost related to sensitivity

Table 5-2 (Cont.)

Method	Measures or Detects:	Applications	Advantages	Limitations
Magnetic particle	Surface and slightly subsurface defects; cracks, seams, porosity, inclusions Permeability variations Extremely sensitive for locating small tight cracks	Ferromagnetic materials; bar, forgings, weldments, extrusions, etc.	Advantage over penetrant is that it indicates subsurface defects, particularly inclusions Relatively fast and low cost May be portable	Alignment of magnetic field is critical Demagnetization of parts required after tests Parts must be cleaned before and after inspection Masking by surface coatings
Magnetic field	Cracks Wall thickness Nonmagnetic coating thickness on steel	Ferromagnetic materials Inspection of coatings on steel Wall thickness of nonmagnetic materials	Measurement of magnetic material properties May be automated Easily detects magnetic objects in nonmagnetic material Portable	Permeability Reference standards required Edge-effect Probe lift-off Loss of accuracy on curved surfaces

Table 5-2 (Cont.)

Method	Measures or Detects:	Applications	Advantages	Limitations
Microwave (300 MHz-300-Ghz)	Cracks, holes, deboned areas, etc., in non-metallic parts Changes in composition degree of cure, moisture content Thickness measurement	Reinforced plastics	Between radio waves and and infrared in the electromagnetic spectrum Portable Contact with part surface not normally required Can be automated	Will not penetrate metals Reference standards required Horn to part spacing critical Part geometry Wave interference Vibration
Penetrants (Dye or fluorescent)	Defects open to surface of parts; (cracks, porosity, seams, laps, etc.) Through-wall leaks	All parts with nonabsorbing surfaces forgings, weldments, castings, etc.)	Low cost Portable Indications may be further examined visually Results easily interpreted	Surface films, such as coatings, scale, and smeared metal may prevent detection of defects Parts must be cleaned before and after inspection Defect must be open to surface Bleed-out from porous surfaces can mask indications of defects

Table 5-2 (Cont.)

Method	Measures or Detects:	Applications	Advantages	Limitations
Radiography (X-rays-film)	Internal defects and variations; porosity; inclusions; cracks; lack of fusion; geometry variations; corrosion thinning	Castings	Permanent records; film	High initial costs
	Density variations	Electrical assemblies	Adjustable energy levels (5 kv-25 mev)	Orientation of linear defects in part may not be favorable
	Thickness, gap and position	Weldments	High sensitivity to density changes	Radiation hazard
	Misassembly	Small, thin, complex wrought products	No couplant required	Depth of defect not indicated
	Misalignment	Nonmetallics	Geometry variations do not affect direction of X-ray beam	Sensitivity decreases with increase in scattered radiation
		Composites		
Ultrasonic (0.1-25MHz)	Internal defects and variations; cracks, lack of fusion, porosity, inclusions, delaminations, lack of bond, texturing	Wrought metals	Most sensitive to cracks	Couplant required
	Thickness	Welds	Test results known immediately	Small, thin, complex parts may be difficult to check
		Brazed joints	Automating and permanent record capability	Reference standards required
		Adhesive-bonded joints	Portable	Trained operators for manual inspection
		Nonmetallics	High penetration capability	Special probes
		In-service parts		

Source: NACE, 1980.

visual inspection is required to detect deteriorating areas before they develop into serious problems. This can be accomplished by careful picking and scraping of suspected spots to locate surface corrosion. Rust scale should be removed wherever detected.

Figures 5-1 and 5-2 illustrate specific areas of concern that should be addressed during the visual inspection of aboveground tanks. Table 5-3 provides a legend explaining details noted in Figure 5-1. Visual detection of suspected cracks in the tank shell or welds should be followed by a more detailed assessment. The tank should immediately be taken out of service, and the entire area surrounding the crack sandblasted. The extent of the crack can then be determined using the magnetic particle method (for metallic tanks) or the dye penetrant method. If the magnetic particle method is employed on a tank that is in service, the magnetic field should be energized with a magnet rather than direct current. Repair, as necessary, should be undertaken immediately.

When a tank is out of service, wall thickness measurements should be taken to establish and monitor the rate of internal shell corrosion. Obvious pitting can be measured with hand calipers, but overall thickness measurement is typically performed using ultrasonic techniques. Ultrasonic readings should be taken at several intervals along the height of the tank to compensate for the varying height of stored liquid in the tank. Readings obtained from the outside of the tank should be compared with readings taken from inside the tank to insure consistency and accuracy.

5.2.2 Underground Tanks

After proper siting and tank type selection, the next step in insuring the integrity of underground tanks is to employ the proper installation procedures, such as those discussed in Section 3.3.1. Unless underground tanks are installed in a vault, external inspections of the tanks are not feasible except by unearthing the tank. Internal inspection may be possible if the tank is equipped with a manhole. Leak detection is the primary way problems are identified.

NFPA Bulletin No. 329, "Recommended Practice for Handling Underground Leakage of Flammable and Combustible Liquids," provides criteria for so-called "final testing" of underground storage tanks. The bulletin states that "the test will conclusively determine whether or not an underground storage and liquid handling system is leaking. Any testing devices used for the final test shall be capable of detecting leaks as small as .05 gallons in one hour, adjusted for variables...." (Note: Proposed new terminology in a draft NFPA #329 will change this definition to "Precision Testing").

Air testing of buried tanks, once a common practice, is not recommended. Not only is it an extremely unreliable method of detecting existing leaks, particularly those below the liquid level, but it can worsen leaks or create leaks that did not previously exist. There is also a serious danger of causing tank rupture. Air testing is an acceptable method of leak detection only when it is used as a

Table 5-3

TYPICAL ABOVEGROUND TANK SYSTEM - AREAS OF CONCERN

Item	Requirement
A	Tank fill valve should be in the closed position and locked when not in use.
B	The gate valve used for emptying the diked containment area should be of the hand-operated variety only and should be closed and locked at all times.
C	All valves should be inspected for signs of leakage or deterioration.
D, E	Inlet and outlet piping, as well as tank flanges should be checked for leakage and to insure that adequate support is provided
F, G	Automated fill control and discharge control equipment should be checked to see that it is operating properly.
H	The tank shell surface should be visually inspected for areas of rust, or other deterioration. Particular attention should be paid to peeling area, welds and seams.
I	The ground surface inside the diked area should be checked for obvious signs of leakage or spillage.
J	The liquid level sensing device should be checked to insure that there is adequate freeboard.
K	External stairways and walkways should be checked to insure that they are unobstructed and sound.
L	The oil/water seperator should be checked for adequate feeeboard and to insure that it is operating properly.

Source: Ecology and Environment, Inc., 1983

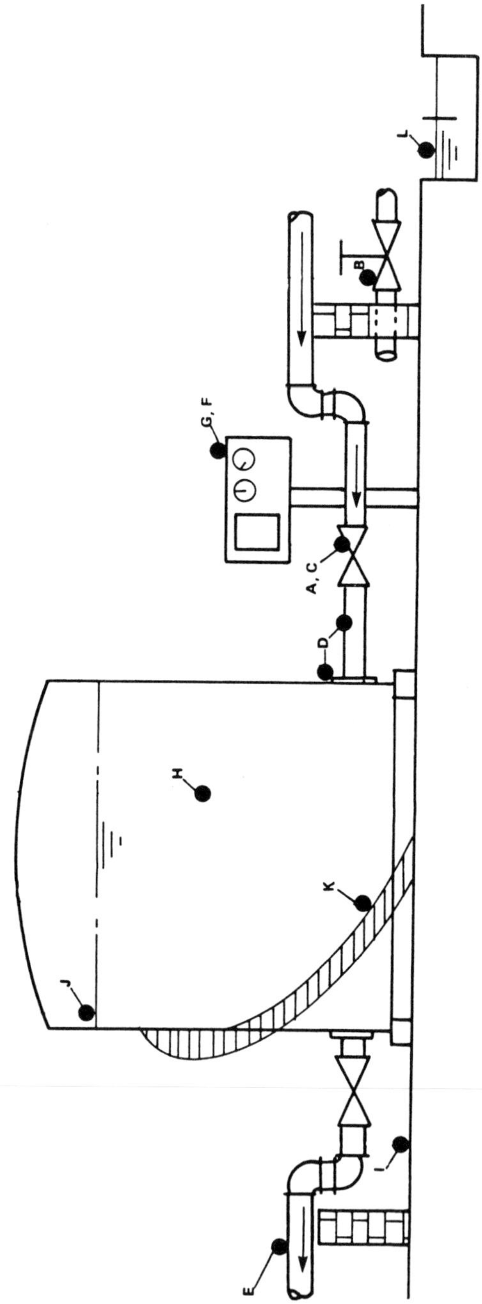

SOURCE: Ecology and Environment, Inc., 1983.

Figure 5–1 AREAS OF CONCERN IN TYPICAL ABOVEGROUND VERTICAL TANK SYSTEM (See Table 5–3 for legend)

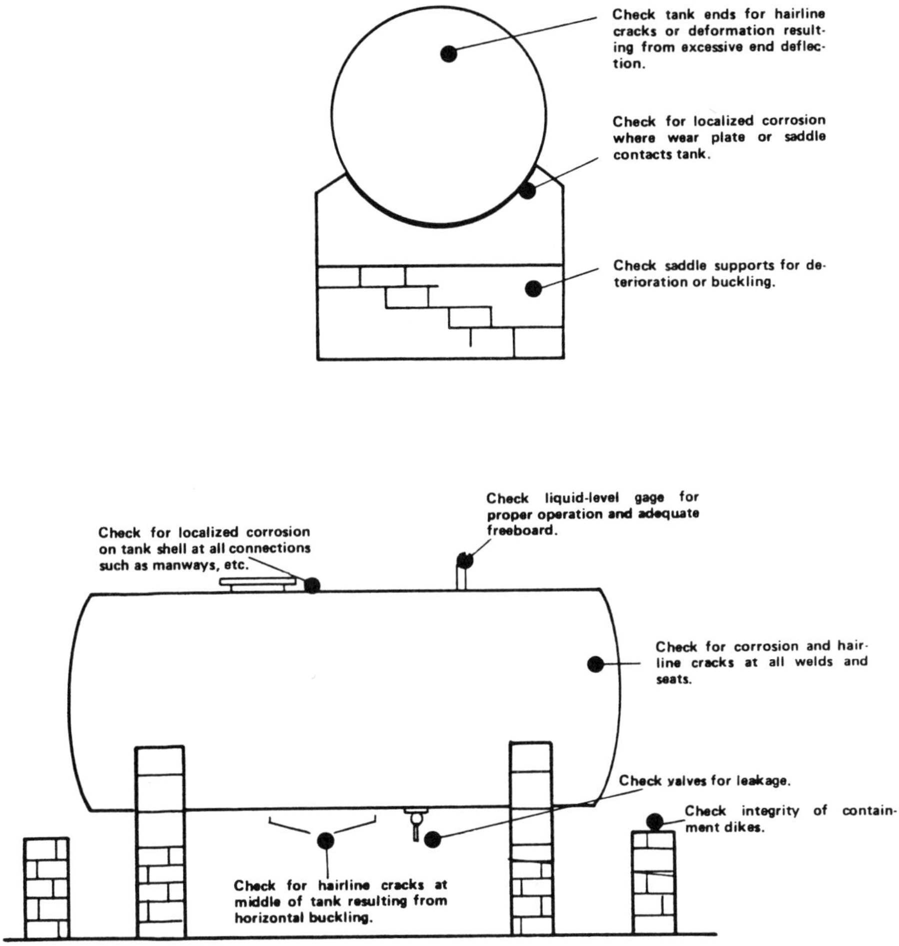

SOURCE: Ecology and Environment, Inc., 1983.

Figure 5–2 AREAS OF CONCERN IN TYPICAL HORIZONTAL TANK SYSTEM

check test on empty tanks prior to installation, where all joints, seams, and welds can be soaped and observed for the presence of air bubbles.

Acceptable test methods for underground tanks fall into at least three generic categories: hydrostatic tests, liquid-level measurement tests, and sonic tests. The commonly used hydrostatic test involves adding a standpipe extension to the fillpipe of a full underground tank, thereby creating a hydrostatic head on the contents of the tank. This added head must be kept small to avoid excessive overpressuring of the tank. Changes in the liquid level in the standpipe magnify volumetric changes in the tank itself, and thus the rate of change reflects "in-leaks" or "out-leaks" in the tank. A problem arises in that the liquid level is also affected by volumetric changes resulting from the thermal expansion or contraction of the product, as well as tank-end deflection caused by the slight standpipe overpressuring. An accurate result is thus best obtained by techniques which compensate for these factors. The Heath Petro-tite Tank and Line Testing System (Kent-Moore test) is one technique which accomplishes this, while detecting leaks as small as the requisite NFPA criteria of .05 gal/hr.

Liquid-level measurement techniques involve such principles as buoyant force or manometer pressure differentials. The merits of these techniques must be assessed on a case-by-case basis, giving particular attention to the technique's consistency with NFPA requirements, and its ability to compensate for thermal expansion and contraction. Among the liquid-level measurement techniques are the J-tube manometer test, the Arco HTC Storage Tank Leak Tester, laser beam leak detection, and the Sunmark Leak Lokator. Each of these tests is based on different principles and has its own distinct advantages.

The J-tube manometer test utilizes a manometer-type instrument composed of a narrow-diameter J-tube indicator attached to a large-diameter reservoir tube. When placed in an underground tank, any change in the tank level will cause a corresponding displacement of fluid in the indicator tube. Over a sufficient time period, the test can detect level changes as small as 0.02 inches. However, product temperature changes as low as 1°F can negate the results, and it cannot detect leaks above the product level in the tank.

The Arco HTC Storage Tank Leak Tester uses a float positioned at a calculated depth, and a light-sensing detector attached to a support rod. The amount of liquid in the detector light path changes in proportion to changes in liquid volume in the tank. This in turn changes the amount of light seen by the detector's photocell, which causes a corresponding voltage change. The recorded voltage changes are then used to calculate the liquid volume change. The test detects leaks less than 0.05 gal/hr, but requires frequent recalibration during the test. Additionally, the test is limited to tanks which are approximately 75% full.

The Sunmark Leak Lokator consists of a sensor suspended from an analytical balance and partially submerged in the tank liquid. Filled with the same liquid as the tank, the sensor's buoyancy changes as the tank liquid level changes. The test is sensitive to 0.03 gal/hr, compensates for temperature and pressure, and can distinguish between tank or piping leaks.

The sonic measurement technique is based upon the principle that when the headspace pressure within the tank is reduced by an amount slightly in excess of the equivalent head of fuel in the tank, air will be drawn in through any leak, with the attendant formation of bubbles on the interior tank wall. As these bubbles detach from the wall they emit a distinct sound that is readily distinguishable from background noise. This technique is thus unaffected by temperature or volumetric changes within the tank. On the other hand, the technique detects only the existence of a leak and not the rate of leakage. It has yet to receive widespread acceptance, and is currently only applicable to gasoline.

A format for documenting leak test results is shown in Figure 5-3. Although the specific information to be recorded will vary with the type of test used, the format illustrates the type of data that should be recorded.

An indirect means of underground tank inspection is to maintain inventory control records for individual tanks. Thorough records of transfer of product to and from a tank will reveal losses resulting from unseen leaks.

Monitoring wells are appropriate means of leak detection at facilities storing large volumes of product underground. The wells are usually installed to detect the presence or investigate the movement of liquid or gas in the ground. As such, they may be qualitative in nature, using contaminant sensing devices to determine presence or absence of contaminants, or they may be quantitative, requiring periodic sample collection and analysis. Wells may be used to detect vapors in unsaturated, permeable soils, or they may be used to detect or determine the movement of leaked substances in the groundwater.

Placement of qualitative wells should be at those locations most likely to provide an early indication of leakage. Quantitative groundwater monitoring well placement should be such that at least one well is situated hydrologically upgradient of the underground tanks, and at least two wells are situated downgradient. In this manner a comparison can be made of groundwater quality upgradient and downgradient of the storage facilities. The wells may be placed such that monitoring at several depths is provided. Leaks can be detected by monitoring groundwater samples for the compounds that are in storage. Monitoring frequency should be quarterly, at a minimum, but more frequently if there is a suspicion that leakage is occurring.

LEAK TESTING LOG

FACILITY OWNER: ____________ CAPACITY: ____________

ADDRESS: ____________ VOLUME IN TANK: ____________

TANK: ____________ PRODUCT ADDED TO FILL TESTER: ____________

CONTENTS: ____________ TOTAL QUANTITY IN TANK: ____________

Time	Description of Procedures	Volume or Level	Change in Volume or Level	Temperature Change	Net Volume Change	
					This Reading	Cumulative

Test Results/Certification: ____________

____

____

____
Signature of Tester

Organization: ____________

Address: ____________

____

SOURCE: Ecology and Environment, Inc., 1983.

Figure 5–3 LEAK TESTING LOG

5.2.3 Reinforced Plastic (RP) Tanks

Aboveground RP tanks should be visually inspected for flexural cracks after delivery as well as during service. Structural failure resulting from overpressuring will be manifested by interior longitudinal cracking in horizontal tanks and by vertical cracking in vertical tanks. Suspected cracks or other surface discontinuities should be further investigated using the dye penetrant method.

Some attention should be given to the interior laminate for signs of decomposition resulting from chemical attack. If accessible, attention should also be focused on dish tank heads or ends, nozzles, and gussets, all of which are potential weak points on RP tanks.

New tanks, underground as well as aboveground, should be hydrostatically tested prior to installation. If possible, purchasers of RP tanks should require verification of hydrostatic integrity from the manufacturer prior to delivery.

Underground RP tanks may be subjected to the leak detection methods discussed in Section 5.2.2.

5.2.4 Tank Liners

Inspection of tank liners should commence during the installation process. The new liner should be of even thickness and texture, and applied in such a manner that a complete bond is assured. It is especially critical that equipment, particularly ladders, used to apply a liner not introduce punctures or "pinholes" to the fresh surfaces. Adequate cushioning should be used to cover all sharp or blunt objects placed on the liner surface during installation or inspection activities.

Once a liner is placed in service, inspections should be made during the scheduled "down" times for a tank. The minimum inspection frequency for liners is annually, with semiannual or quarterly checks recommended under particularly corrosive conditions.

For any liner, bulging, blistering, and spalling are signs of possible leakage. Lead liners, which are often used over steel for very corrosive materials, should be lightly scraped with a knife or other hand tool to remove the thin dark outer layer of lead oxide. Cracks or other defects will manifest themselves as contrasts to the bright lead thus revealed.

Nickel, monel, alloy steel, or other metal liners should be visually checked for cracks in the joints or seams. Suspect areas can be further checked by the dye penetrant method. Note that the magnetic-particle method cannot be used if the liner material is non-magnetic.

In the case of rubber, glass, organic, or inorganic liners, holes will be readily evidenced by blistering or bulging around them. Visual inspection should be followed up by a spark test which involves passing a high-voltage, low-current electrode over the nonconductive lining, with the other end of the circuit attached to the steel of the tank. An electrical arc will form between the electrode and the steel tank through any holes in the lining. This test should only be performed in well ventilated tanks free of flammable or combustible vapors. This test is described in more detail in Chapter IV of API's "Guide for Inspection of Refinery Equipment." The main limitation of spark testing is the potential damage to lining materials if the applied voltage is too high. Careful attention to spark testing equipment specifications and manufacturers' restrictions should prevent this possibility.

Concrete liners should be visually inspected for rust spots as evidence of cracks and leaks. Light hammer tapping will reveal any loss of bond between concrete and shell through the difference in sound from bonded areas.

5.2.5 Valves

Frequent visual inspections should be made of all valves associated with liquid storage tanks. Attention should be focused on the valve connections and packing glands. The first attempt at stopping a leak would be to tighten the valve flange bolts or packing gland. Threaded valve connections should be checked for corrosion and for stripped or crossed threads.

While a storage tank is out of service, valves should be dismantled to examine previously inaccessible internal parts. Body thickness measurements should be taken with calipers, especially if there is evidence of corrosion. Valve bodies should also be checked for internal cracking.

All seating surfaces should be checked for smoothness and snug fit. The bottom seats of gate valves should be checked for deterioration resulting from turbulence. The diaphragm in diaphragm valves should be checked for its integrity and replaced as necessary. Valve lining material should also be inspected for excessive wear or points of corrosion.

After a valve is reassembled it should be pneumatically or hydrostatically tested. If the pneumatic method is chosen, a soapy solution should be applied around the stem and seating surfaces to detect leaks.

Specific critical areas of concern for common valve types are illustrated in Figures 5-4 through 5-10.

5.2.6 Tank Appurtenances

All tank appurtenances and support systems should be visually checked to insure their integrity and that they are functioning properly. Included should be:

- Stairways and walkways - Check for structural stabilty and for missing treads, rungs, handrails, etc.

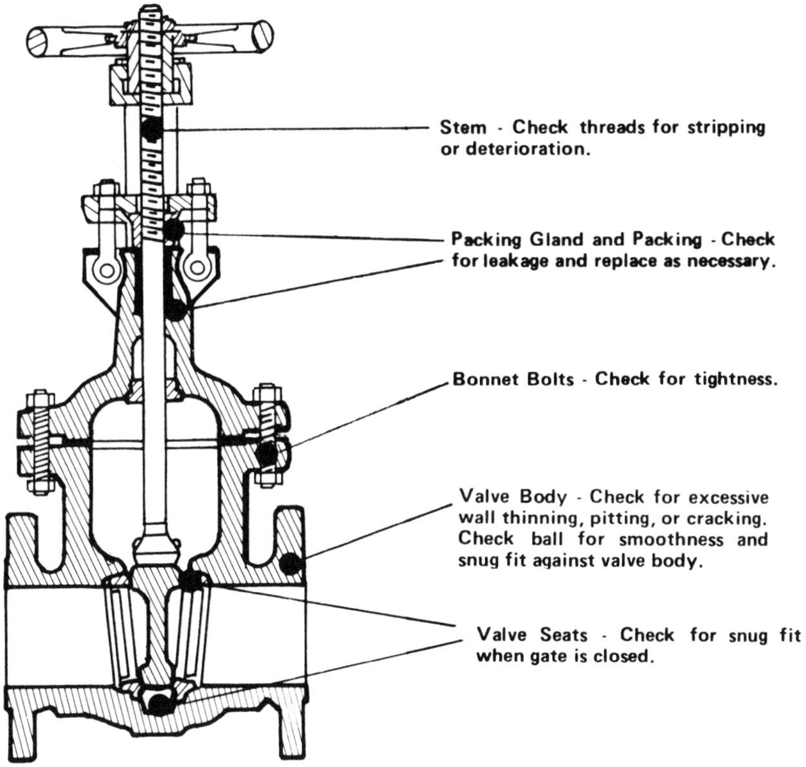

SOURCE: British Valve Manufacturers Association, 1966.

Figure 5–4 CRITICAL AREAS OF GATE VALVE

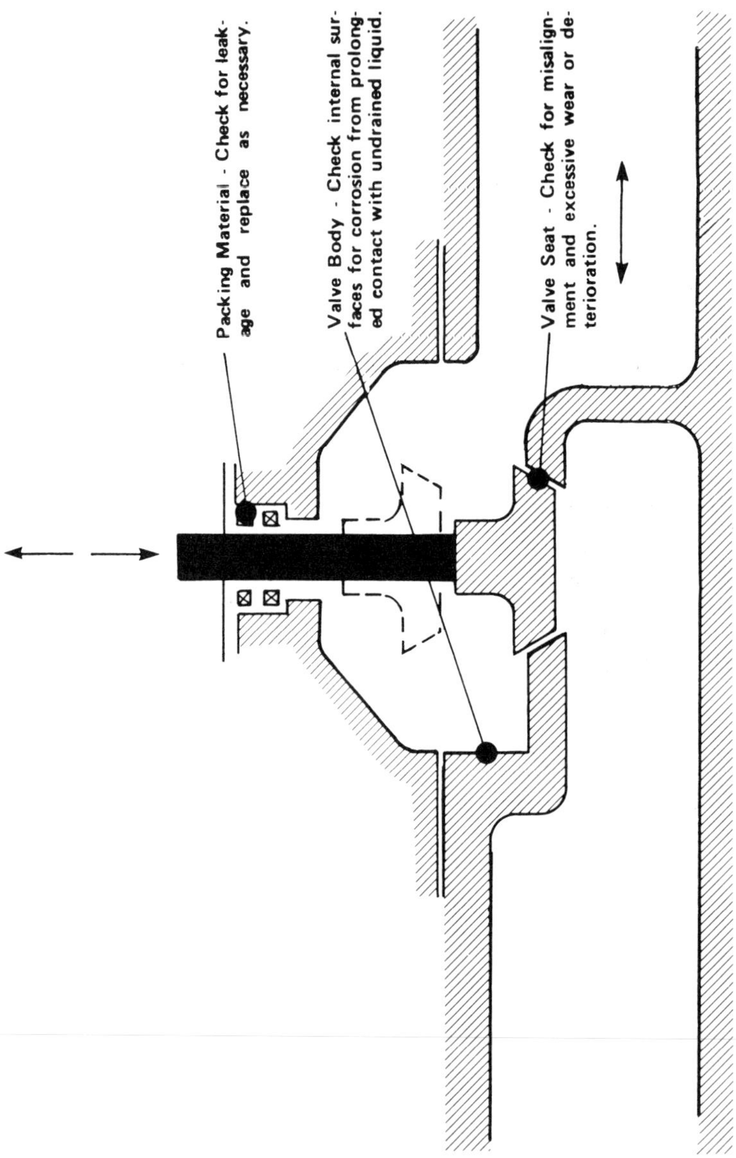

Figure 5–5 CRITICAL AREAS OF GLOBE VALVE

SOURCE: Perry and Chilton, 1973.

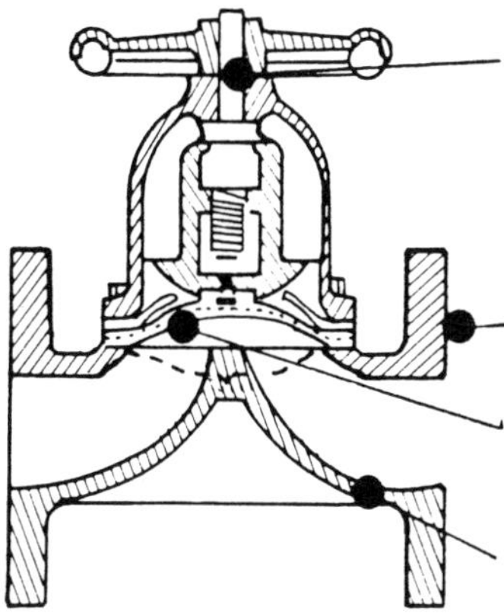

SOURCE: British Valve Manufacturers Association, 1966.

Figure 5–6 CRITICAL AREAS OF DIAPHRAGM VALVE

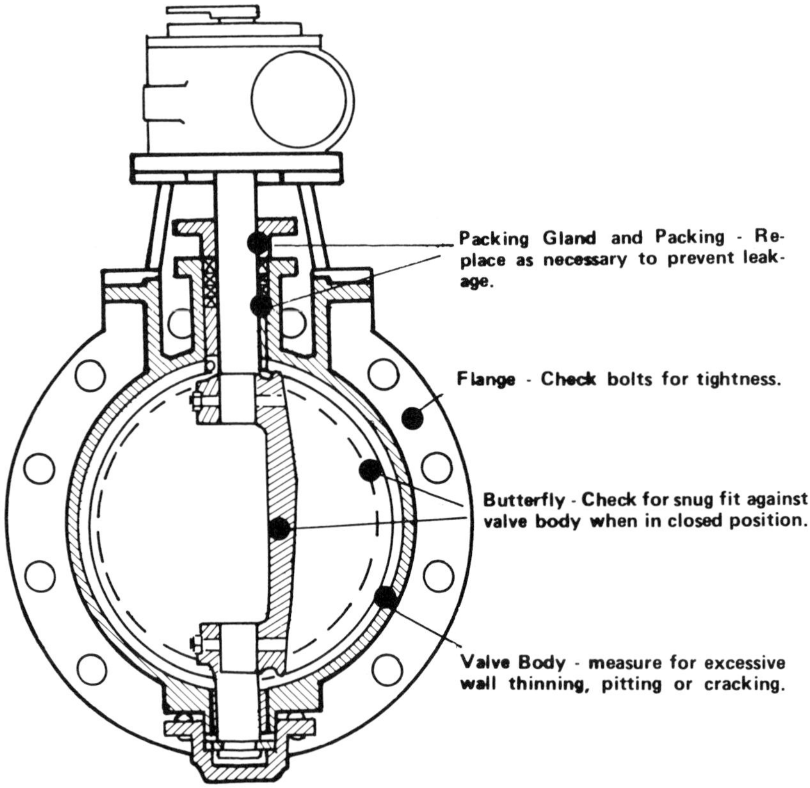

SOURCE: British Valve Manufacturers Association, 1966.

Figure 5–7 CRITICAL AREAS OF BUTTERFLY VALVE

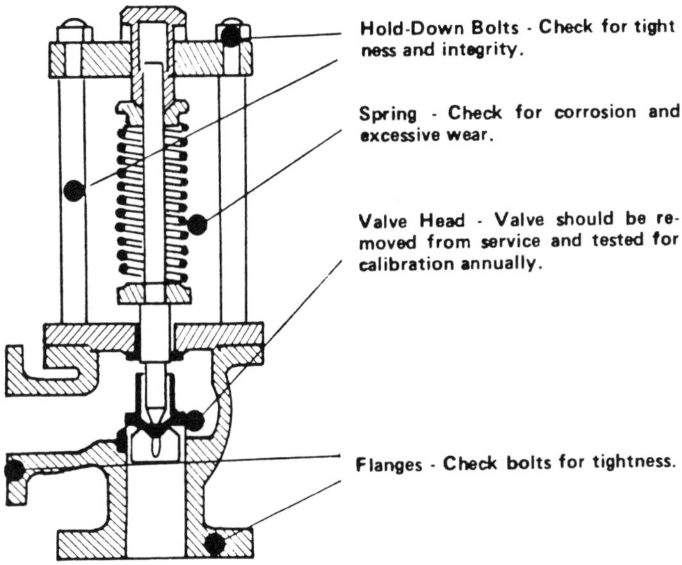

SOURCE: British Valve Manufacturers Association, 1966.

Figure 5–8 CRITICAL AREAS OF SAFETY (PRESSURE RELIEF) VALVE

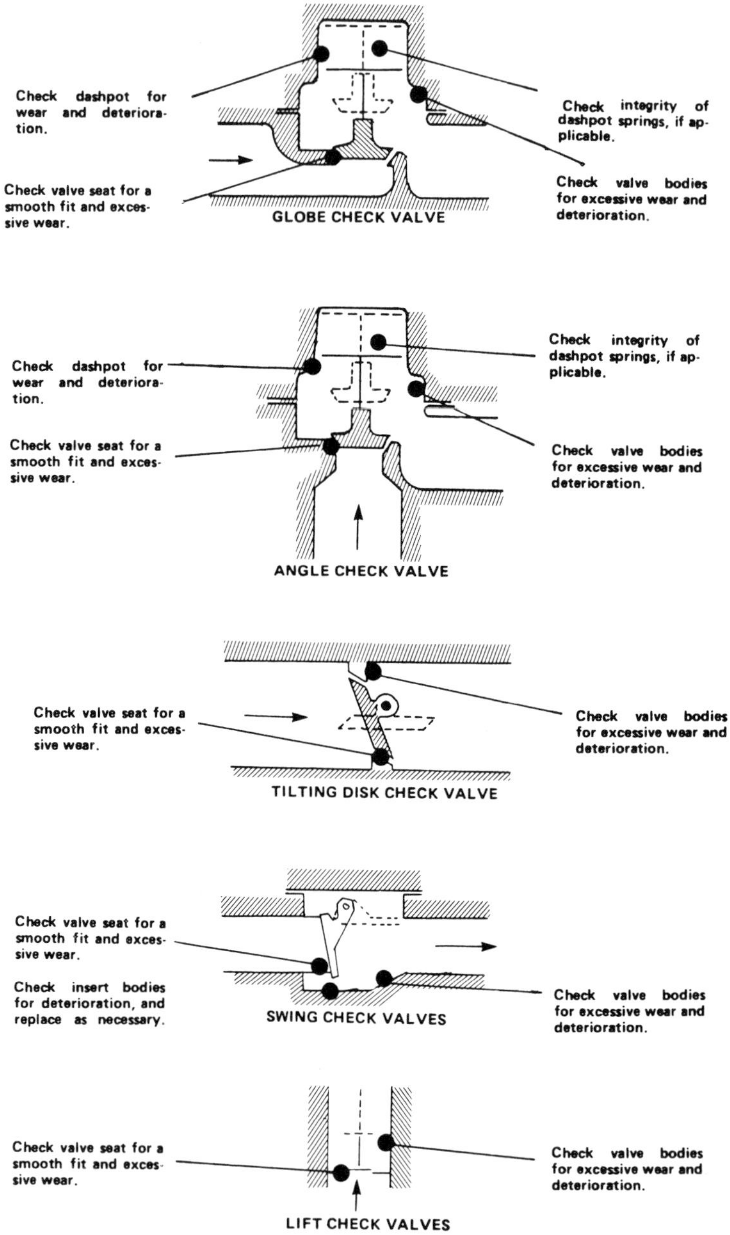

SOURCE: Perry and Chilton, 1973.

Figure 5–9 TYPES AND CRITICAL AREAS OF CHECK VALVES

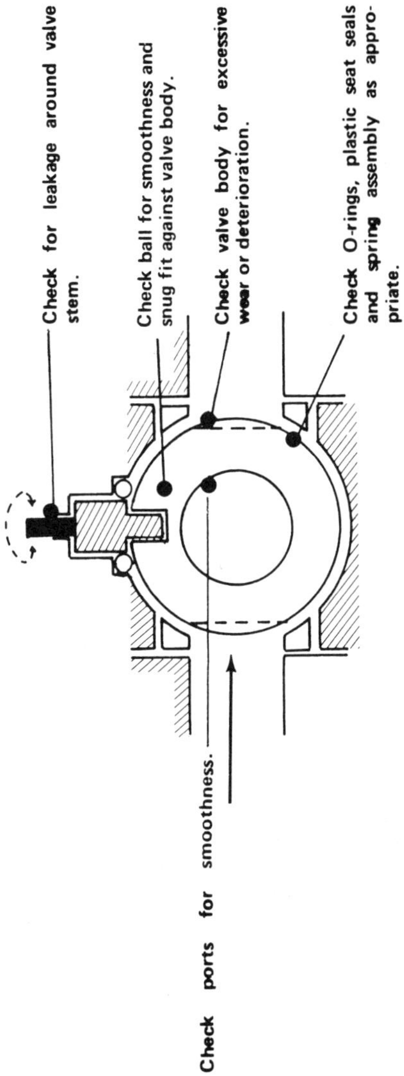

SOURCE: Perry and Chilton, 1973.

Figure 5–10 CRITICAL AREAS OF BALL VALVE

- Flame arrestors - These should be opened at intervals to inspect for cleanliness and corrosion. Remove any clogged debris.

- Dikes and Berms - Check for excessive erosion of earthen structures and general deterioration of man-made materials. Measurements should be taken and compared to original design dimensions.

- Ground Connection - Check visually for corrosion at point of contact with ground and at the mechanical connection to tank. The resistance of the ground should be less than 25 ohms.

- Pressure Vacuum Vents - Check to see that they are not plugged and that all moving parts are free.

- Liquid Level Gaging Equipment - Check for corrosion or cracks and to insure that no liquid is present in the mechanism.

- Floating Roof Water Drains - Check for any breakage or blockage to avoid ponding and collapse. Roof-to-shell seals should be inspected visually for their integrity.

- Electrical devices - Check for frayed or bare wires and connections, proper circuit protection, and accumulation of flammable debris. Electrical devices near flammable material should be approved for use in a flammable materials environment.

All pipe connections should be inspected visually, especially at ground level, for corrosion or distortion. Pipe supports should be checked for structural stability. Ultrasonic readings can be taken to check for thickness reduction from corrosion or pitting.

5.2.7 Foundations

Foundation ringwalls and concrete base curbing should be inspected visually for evidence of general deterioration. Cracks or decay should be repaired immediately both for structural integrity and to prevent precipitation or other liquids from accumulating under the tank. Foundation settlement should be checked against a known reference point, using a surveyor's level.

General support structures, such as piers, columns, legs, and stands, should be checked visually for their integrity, as well as with calipers to detect excessive corrosion. Anchor bolts should be checked for their integrity and tightness by striking them with a hammer or similar instrument. Distortion of anchor bolts or columns is an indicator of excessive settling. The welds along the angle iron at the intersection of the shell and tank botom should also be inspected. Figure 5-11 illustrates specific areas of concern for the inspection of foundations.

In the case of aboveground tanks resting directly on a soil foundation, it is usually impossible to visually detect leaks in the

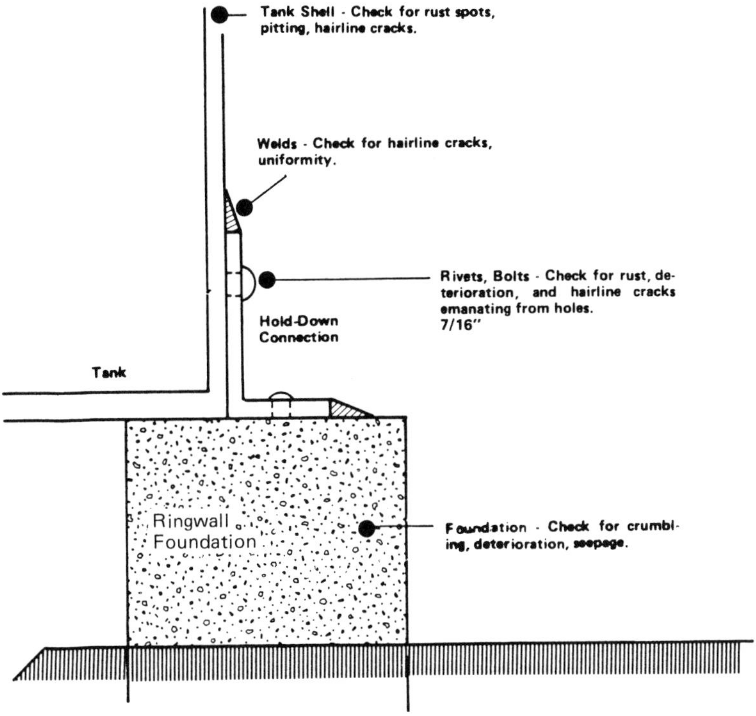

SOURCE: Ecology and Environment, Inc., 1983.

Figure 5–11 AREAS OF CONCERN IN A TYPICAL TANK FOUNDATION

tank bottom plate. One mechanical method would be to place a temporary clay dam or seal (if one does not already exist) around the base of the tank and inject air underneath the tank at a pressure equal to no more than three inches of water. Leaks will evidence themselves as air bubbles when a soap solution is applied to the interior tank bottom. An alternative method is to soap suspected areas of the tank bottom and then apply a gasketed vacuum box. As a vacuum is drawn within the box, leaks will, again, evidence themselves as air bubbles. The vacuum box technique can be applied to any tank surface.

5.2.8 Cathodic and Anodic Protection Systems

Storage systems equipped with cathodic or anodic corrosion controls require periodic inspection of those controls if they are to provide the long-term protection desired. For galvanic anode systems, there should be at least annual measurements of tank-to-soil potential and anode output. Items to check in all galvanic systems include broken wires, broken or shorted insulators, or loss of coatings. Impressed current electrode systems require inspection to detect potential failure due to power interruption, malfunction of rectifiers, deterioriation of anodes, and broken wires. Rectifier output should be monitored monthly with a voltage or amperage indicator, and adjusted as needed. Tank-to-soil potential measurements should be made at least annually to determine if rectifier adjustments are needed to maintain adequate corrosion protection.

5.3 FREQUENCY OF INSPECTION

The intervals at which various inspection tasks should be performed is largely site-specific. No two storage situations are exactly alike, and it is because of the dissimilarities that it is difficult to set rigid inspection frequencies. The frequencies should be set on a case-by-case basis, considering:

- The chemical nature of the material being stored;
- Known or expected corrosion rates;
- The corrosion allowance inherent in the tank wall thickness;
- Previously observed conditions; and
- Tank location.

Many external components of a storage system can be easily and routinely inspected through visual observations or simple mechanical checks. These general elements are listed in Table 5-4 as daily or weekly tasks. Despite the routine nature of these checks, a conscientious program of external inspection should be adequate to detect preliminary corrosion, leaks, excessive settlement, and improper functioning of vents, pressure-relief valves, gaging devices, and other appurtenances.

A more thorough inspection of external tank surfaces, welds, rivets, and foundations should be undertaken on a monthly basis. Upon close inspection special attention should be paid to the critical areas discussed and illustrated in Section 5.2.

Table 5-4

MINIMUM INSPECTION TASKS AND FREQUENCIES

Frequency	Task
Daily	Visually check valve stems and flanges for leakage
	Visually check piping for misalignment, bending, or leakage with particular attention to tees, couplings, elbows, and connections
	Inspect ground surface around vertical and horizontal tanks for signs of leakage
	Check discharge and fill control equipment before product is transferred to insure that it is functioning properly
	Check liquid level in the tank before product is added to insure adequate capacity
	Check gate valve from diked area to insure that it is closed and locked
	Check walkways and stairways for obstructions
	Check and record inventory of tank contents
Weekly	Check liquid level gaging equipment to insure that it is functioning properly
	Check roof drains for obstructions
	Check vents and pressure-relief devices for obstructions
	Check grounding lines and connections for integrity
	Check stairways for damaged rungs or handrails
	Check containment dike or berm for integrity
	Does oil/water separator or equivalent require pumping
	Check separator discharge for clarity
	Does diked area require drainage
	Check fire extinguishing equipment
Monthly	Inspect all exterior tank surfaces, welds, rivets/bolts, foundation
	Check impressed current rectifiers
	Inventory all spill control and other emergency response equipment
Quarterly	Non-destructive thickness testing of piping and valves

Table 5-4 (Cont.)

Frequency	Task
Semi-Annually (at scheduled down-time)	Thickness testing for shell walls Inspection of liners Leak testing of foundation Leak testing of underground tanks assembly
Annually (at scheduled down-time)	Test structural stability of support structures for elevated tanks and test pressure relief valves for calibration Measure tank-to-soil potential

Source: Ecology and Environment, Inc., 1983.

The most detailed inspection tasks are those which require disruption of service, access to the interior of the tank, or disassembly of components. Necessarily, these tasks are undertaken on a less frequent basis, such as quarterly, semi-annually, or annually, as outlined in Table 5-4. These types of tasks generally require the use of the nondestructive test methodologies detailed in Section 5.1.

5.4 INSPECTION CHECKLISTS

The inspection process should be somewhat structured at least to the extent that there are minimum inspection tasks (outlined in Section 5.3). The use of formal checklists, logs, and report forms will insure thorough and complete attention to all required inspection tasks, as well as provide a hard-copy record of conditions which could warrant corrective or maintenance activities. In this manner, recurring or developing problems would be readily identified and remedied.

Sample inspection checklists are illustrated in Figures 5-12 through 5-14. The daily inspection checklist should be used to record the routine yet important observations that should be made on a day-to-day basis. The weekly checklist should cover those areas that ordinarily receive less frequent attention but are fundamental to the safe operation of a tank or tank system. The monthly inspection log should yield a more detailed assessment of the structural integrity of the tank, and should form the basis for any required corrective action. It should be stressed that individual checklists must be developed on a facility-by-facility basis to adequately address the range of conditions encountered in the field.

The thickness testing log (Figure 5-15) provides a means for tracking shell, valve, or piping wall deterioration. The results of non-destructive thickness testing performed as part of the schedule outlined in Section 5.3, or on an as-needed basis, should be recorded in this log. This log may be used to develop the graphical techniques outlined in Section 5.5, Criteria For Correction Action. Note that the last column in the log provides for the computation of safe loading parameters for the tank in the event that wall-thinning has appreciably reduced the strength of the tank.

All completed forms should be kept in an ongoing file for each tank. This file may correspond with the SPCC records recommended in Section 7.

These logs and checklists should be modified to correspond to individual tank characteristics or conditions.

5.5 RATIONALE FOR CORRECTIVE ACTION

As discussed in Section 3.7, one of the purposes of a hazards and risk analysis of a proposed hazardous materials storage facility is the early determination of corrective design actions that will eliminate or minimize the identified critical hazards. By extension, the logical principles of hazard or risk analysis may be applied to determine appropriate corrective actions, thereby placing the greatest priority on correcting those hazards identified as having the greatest

Daily Inspection Checklist

TANK # ______________________ PAGE ______ OF ______

LOCATION ______________________ INSPECTED BY ______________

FILL LEVEL ______________________ DATE ______________

CONTENTS ______________________

Inspection Tank	Accept-able	Unaccept-able (Specify)	Recommended Corrective Action	Referred To
Is fill valve locked and closed?				
Is dike gate valve locked and closed?				
Condition of valve # ___ # ___ # ___				
Condition of inlet piping				
Condition of outlet piping				
Fill control equipment functioning?				
Discharge control equipment functioning?				
Visual check of tank shell integrity				
Evidence of leakage on ground?				
Is there adequate freeboard in tank				
Are stairways and walkways unobstructed				

SOURCE: Ecology and Environment, Inc., 1983.

Figure 5–12 DAILY INSPECTION CHECKLIST

WEEKLY INSPECTION CHECKLIST

TANK # ________________ PAGE ______ OF ______

LOCATION ________________ INSPECTED BY ________________

FILL LEVEL ________________ DATE ________________

CONTENTS ________________

Item	Accept-able	Unaccept-able (specify)	Recommended Corrective Action	Referred To
Is liquid level gaging equipment operating properly?				
Is emergency shut-down system func-tioning properly?				
Are roof vents clear?				
Are roof drains clear?				
Are pressure relief devices free of obstruction?				
Are stairways and handrails in good condition?				
Is containment dike/berm intact?				
Does diked area require damage?				
Is there adequate freeboard in oil-water separator				
Is fire extinguish-ing equipment in place and func-tioning properly?				
Leak detection equipment func-tioning properly?				

SOURCE: Ecology and Environment, Inc., 1983.

Figure 5–13 WEEKLY INSPECTION CHECKLIST

Item	Accept-able	Unaccept-able (specify)	Recommended Corrective Action	Referred To
Check ignition safeguards: Isolated metal objects and fill nozzels grounded?				
Absence of spark promoters from tank interior?				
Inlet flow rate sufficiently limited?				
Ground con-nections secure?				
Metallic shunts intact?				

SOURCE: Ecology and Environment, Inc., 1983.

Figure 5—13 WEEKLY INSPECTION CHECKLIST (Cont.)

MONTHLY TANK INSPECTION LOG

TANK # ________ LOCATION: ________ FILL LEVEL: ________

CONTENTS: ________ INSPECTED BY: ________ DATE: ________ PAGE ___ OF ___

Item	Observations	Recommendations for Corrective Action
Tank shell and roof -Discoloration or flaking of coating -Localized corrosion -Structural damage -Development of hairline cracks -Bulging or cavitation -Deterioration at joints and connections		
Welds -Localized corrosion -Separatin or distortion of welded components -Development of hairline cracks		
Rivets/bolts -Localized corrosion -Loosened components -Missing		
Foundations/supports -Cracking or deterioration of concrete ringwall or support -Uneven settlement -Slippage of tank from foundation or support -Buckling of saddle or vertical supports -Loosened anchor bolts		

SOURCE: Ecology and Environment, Inc.. 1983.

Figure 5–14 MONTHLY TANK INSPECTION LOG

THICKNESS TESTING LOG

TANK # ____________ LOCATION: ________________ CONTENTS: ________________

Date of Inspection	Fill Level	Thickness, Inches																	Tank Certified for Continued Service at Liquid Height h = Liquid Gravity G =
		h_1	h_2	h_3	h_4	h_5	h_6	h_7	h_8	$Roof_1$	$Roof_2$	$Bottom_1$	$Bottom_2$	$Valve_1$	$Valve_2$	$Valve_3$	Inlet Piping	Outlet Piping	

SOURCE: Ecology and Environment, Inc., 1983.

Figure 5–15 THICKNESS TESTING LOG

potential for loss of life, personal injury, environmental damage, or property damage. Conversely, those items identified as having no or minimal hazard potential are given lower priorities. Therefore, establishing priorities for corrective action is based largely on an assessment of the potential consequences arising from failure to correct an identified defect.

The potential consequences of equipment failure are numerous. The possibilities include, but are not limited to, any one or combination of the following:

- Loss of life;
- Loss of property;
- Loss of natural resources;
- Loss or adulteration of product;
- Fire and/or explosion;
- Toxic vapor release; and
- Widescale evacuation of populations.

The extent of hazard posed by any facility will be specific to that facility. Therefore, it is not possible to provide specific recommendations regarding critical failure points, etc. Such details are covered by the design and material specifications used in tank or facility construction, and are illustrated throughout Chapters 3 and 5 of this manual. However, the factors a facility operator must consider in developing maintenance and repair priorities include:

- Age of the facility;
- Materials used in construction;
- Physical/chemical/toxicological properties of the stored material;
- Quantity of material stored;
- Operation of warning or control devices;
- Geologic/hydrologic/topographic properties of the site;
- Design tolerances and specifications;
- Size and proximity of adjacent population;
- Value of the stored materials and facility; and
- Consequences of storage system failure.

The inspection schedules and critical areas described throughout this chapter illustrate how the above factors may be incorporated into a routine maintenance and inspection program. However, these serve as illustrative guidelines only. The facility operator should develop a maintenance and inspection program specific to the condition and needs of the individual facility.

The overall objectives of any maintenance and inspection program are to prevent loss of stored material and avert the consequences thereof. Therefore, in any situation it is advisable that corrective action be undertaken at the earliest detection of problems. For example, cracks, discontinuities, corrosion or rust spots, inoperative vents or other equipment, leaky valves, or eroding berms all warrant immediate repair.

In the case of gradual wall thinning of tanks or valves, the point at which repair or replacement is required is determined by the minimum thickness needed to insure structural integrity. For steel tanks this is generally one-quarter inch or loss of 20% of the original wall thickness. For valves, the minimum thickness is calculated according to ANSI standard B34. Graphs should be used to predict, based on historical thickness measurements, the point at which the safe minimum thickness would be reached. Replacement should take place before this point. An example of the technique is illustrated in Figure 5-16.

Tank or valve liners should be replaced when it is no longer feasible to repair them. Valve diaphragms are readily removable and should be replaced as needed.

The above criteria for repair and replacement should be used as guidelines only. Specific criteria must be based on the specifications of the materials actually used and on conditions at the facility. However, priorities for repair and replacement may be established on the basis of a logical analysis of:

- The defects identified;
- The potential consequences of product loss; and
- The costs, practicality, and effectiveness of the repair.

Figure 5-17 illustrates a logical scheme for establishing corrective action priorities. The scheme is a matrix comparing categories of defects, in decreasing order of severity, with a similar range of consequences. The categories of defects are:

- Actual loss of primary containment, which may be caused by tank wall failure, leakage, spills, etc;

- Imminent loss of primary containment, indicated by cracks, excessive corrosion, wall thinning, etc;

- Potential loss of primary containment, indicated by such conditions as loose fittings, inoperable gaging or warning devices, visible corrosion, etc; and

- Inadequate secondary containment, indicated by such conditions as insufficient freeboard, accumulation of rainwater, breached diking, etc.

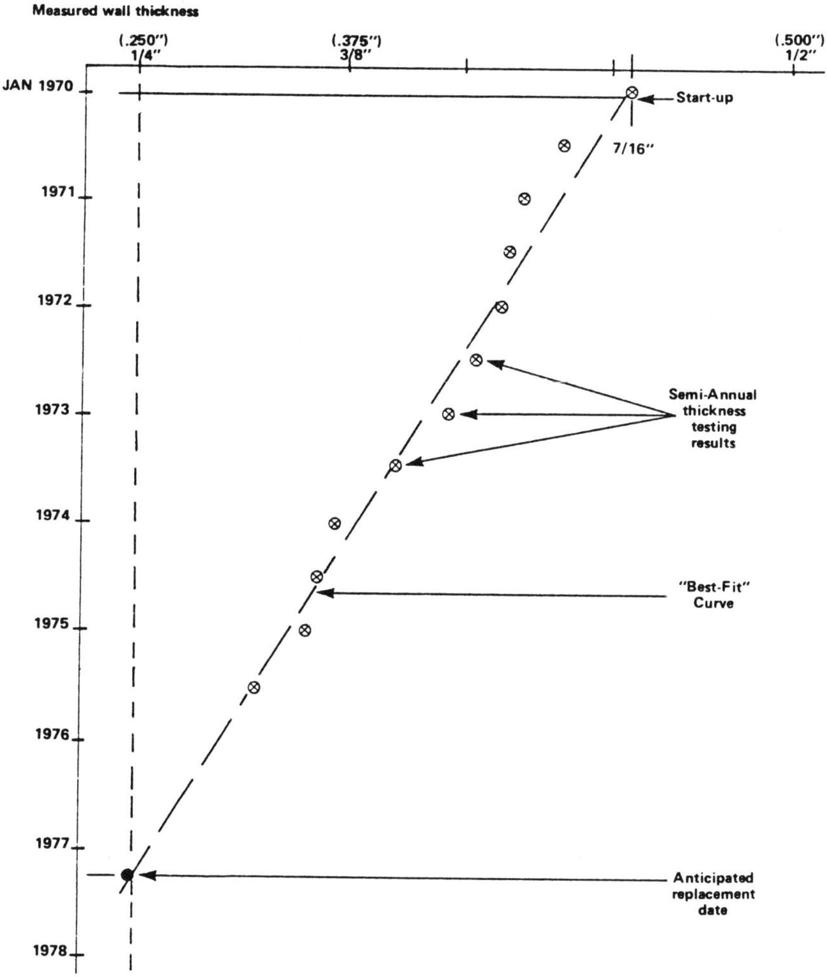

SOURCE: Ecology and Environment, Inc., 1982.

Figure 5–16 GRAPHICAL CALCULATION OF REPLACEMENT DATE FOR TANKS AND VALVES

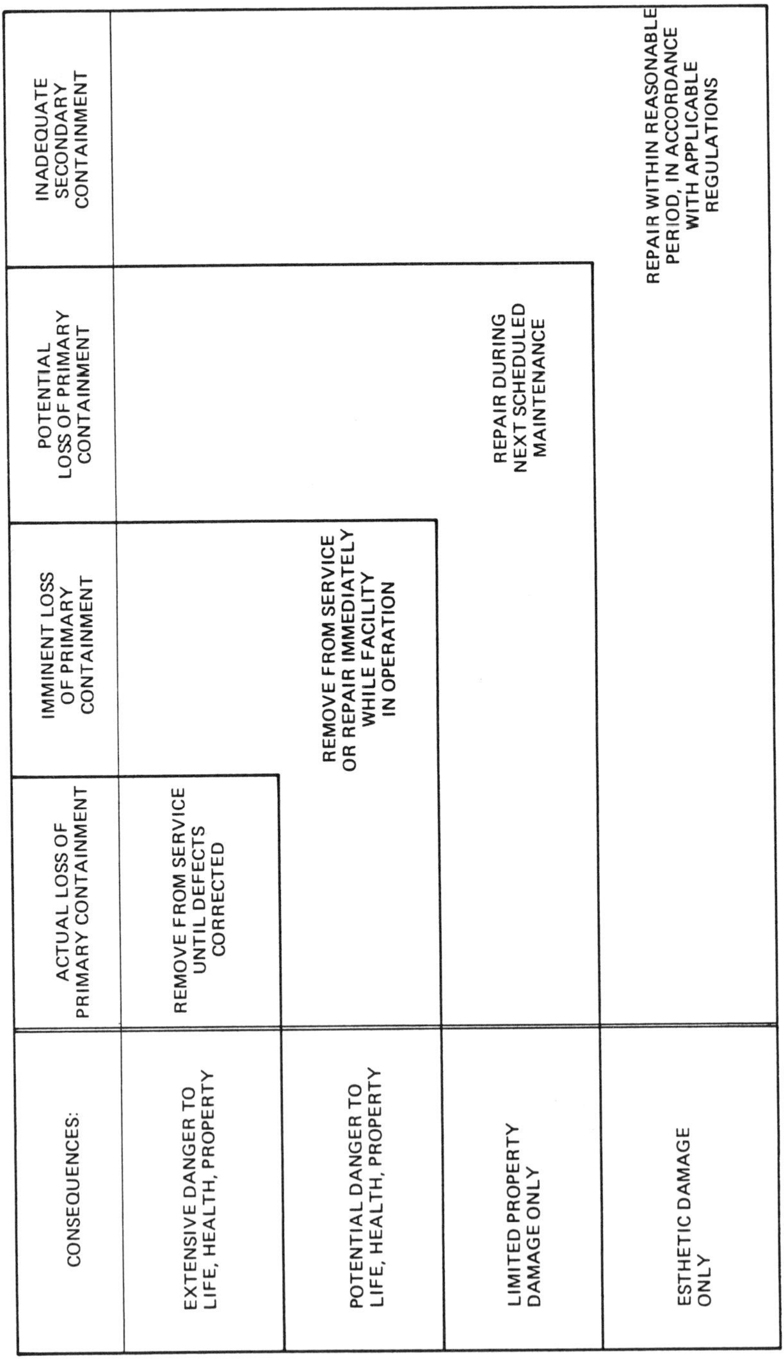

SOURCE: Ecology and Environment, Inc., 1983.

Figure 5—17 ILLUSTRATIVE MATRIX FOR DETERMINING CORRECTIVE ACTION PRIORITIES

Categories of consequences may be described as:

- Extensive danger to life, health, or property, as in a gas release or fire;
- Potential danger to life, health, or property;
- Limited property damage to facility; and
- Esthetic damage.

By matching the defects and consequences in order of severity, one may determine the priorities for corrective action. In this example, these priorities are:

- Removing tank from service until defects are corrected;
- Immediate repair while facility remains in operation;
- Repair during next scheduled maintenance period;
- Repair when manpower and resources permit, within an allotted time period.

The scheme and categories presented in Figure 5-17 are not intended to be all-inclusive, or fully descriptive of the types of defects or consequences that may occur. Nor should it imply that certain classes of defects require only limited corrective action. However, it illustrates a systematic method for analyzing defects in terms of the consequences they may cause, and serves as a guide to plant operators for developing their own repair priorities.

5.6 TANK CLEANING GUIDELINES

Tank cleaning can be extremely dangerous if not performed carefully. Improper removal of even small volumes of solid, liquid or gaseous residues of hazardous chemicals can cause explosions and worker asphyxiation or poisoning. In general, the most important consideration is that the tank cleaning method be compatible with the chemical stored and the tank material, and it is therefore essential that cleaning procedures be well thought out and planned in advance.

Proper procedures and equipment are necessary. This should include maintaining an adequate air supply or respiratory protective equipment for personnel; protective clothing, such as chemical-resistant clothing, rubber boots, gloves, and goggles, and face shields or eye protective equipment; preparation of emergency escape and rescue plans for personnel; appropriate cleaning equipment, which may include steam nozzles, sandblasting equipment, and agitators; safety belts, safety lines and ladders; air monitoring equipment such as combustible gas indicators and oxygen indicators; and first aid. All tank cleaning equipment should be electrically bonded to tank shells during use, and all lighting and electrical equipment used inside or near the tanks should be intrinsically safe or grounded to prevent sparks.

In cleaning tanks, the first step is to drain the tank and lines of their contents. This should utilize equipment suitable for temporary storage of the material, and should be performed with the same degree of attention to chemical compatibility, personnel safety, and ignition safeguards as during normal tank transfer operations. When all liquid and solid contents have been removed, vapors must be purged from the tank. All product lines, steam smothering lines and other lines should be disconnected and blanked prior to purging vapors. Vapor removal can be achieved by steaming, ventilating with air or an inert gas, or displacing with water. However, it is extremely important to employ a method which is compatible with the chemical which was stored. For instance, steaming should not be used on tanks that held water-reactive chemicals, and air should not be used to purge a tank when it could cause the internal vapors to become combustible. In addition, vapors should be removed in a manner that does not pose a respiratory or other threat. A combustible gas and oxygen meter should be used as necessary to determine that a tank has been purged of flammable vapors, or that there is an adequate oxygen supply. Combination instruments which measure both parameters simultaneously are available and are recommended for use.

For tank cleaning with entry, no work should be performed inside the tank unless it has been determined that a safe atmosphere exists in the tank. In general, oxygen content should be greater than 19.5% and no greater than 20.9%, and the combustible gas indicator should read less than 10% of the lower explosive limit (LEL) for the material in question. (Note: acceptable LEL levels may vary in different regulatory jurisdictions. Check applicable regulations to determine acceptable maximum levels.) Oxygen and combustible gas readings should be taken at frequent intervals while work is being performed in the tank. If the atmosphere in the tank is not safe, the workers conducting the tank cleaning should try to ventilate the tank and remove the hazard. If efforts to ventilate the tank are unsuccessful, self-contained breathing apparatus or other appropriate respiratory protection should be made available to and used by persons entering the tank.

Tank cleaning without entry is most easily achieved using steam. Steaming should last at least ten minutes to insure that the entire surface of the tank has been heated to near the boiling point of water. Following steam treatment, the tank should be washed with hot water and allowed to overflow to remove any solid debris. However, whenever a tank is overflowed, care must be taken to avoid any contamination of off-site water sources. If steaming does not sufficiently clean the tank, a chemical cleaning solution can be introduced into the tank. When hot solutions are used, they should be maintained at 170°F to 190°F. Cold solutions can be used in severe cases, but as with any chemical solution, these should only be used after it has been determined that they are compatible with the tank material. For example, if cleaned with caustic solution, aluminum- and zinc-coated tanks can generate hydrogen. Chunks of solid material which cannot be removed chemically can be removed by tumbling a chain inside the tank when flammability is not a danger. Following cleaning, all tanks should be tested with a combustible gas indicator before any welding, cutting, burning, or other spark-producing operation is performed.

Additional guidelines on tank cleaning can be obtained from chemical suppliers; cleaning contractors; API Standard RP2015, "Cleaning Petroleum Storage Tanks"; and NFPA Standard No. 327, "Cleaning and Safeguarding Small Tanks and Containers."

5.7 TANK CLOSURE

Proper attention must be paid to the permanent or temporary closure of hazardous materials storage tanks for many reasons. These include prevention of spills or leaks of any remaining contents; minimizing or eliminating the possibility of residual vapor explosion or fire; prevention of accidents or illegal access to the tank; and insuring the appropriate reuse of the tank.

Temporary closure is generally applicable to structurally sound tanks intended to be put back into service within two years, or scheduled for permanent closure within 90 days. Closure is effected by removing all the contents of a tank and filling it with water and corrosion inhibitor. If the tank was used to store flammable materials, sufficient product could remain in the tank to provide a saturated vapor space. All fill and draw-off lines should be capped with concrete, and all vent lines should be left open.

Permanent closure may be accomplished by either abandonment in place or by removal. Abandonment in place is used whenever the age and salvage value of the tank does not justify removal costs, and when future use of the site would not be affected. The tank should then be completely emptied of all liquids and sludges, and all vapors should be removed. The tank should be thoroughly cleaned and filled with an inert solid such as sand, gravel, or concrete. All lines to and from the tank, and any access points should be capped. Aboveground tanks should be securely anchored in place. Tank removal is preferable to abandonment for tanks with sufficient salvage value, or if future uses of the site so require. As with abandoned tanks, all liquid, solid, and vapor contents should be removed, and the tank thoroughly cleaned. All connections to the tank should be removed and temporarily plugged while the tank is transported from the site. If the tank is to be reused, care should be taken to properly clean and prepare it for its future use. The tank should not be used for storage of a chemical not compatible with the previous contents, nor should a chemical storage tank be reused to store food products for human consumption. If the tank is to be disposed of, it should be dismantled or otherwise rendered unusable.

BIBLIOGRAPHY

American Concrete Institute, 1976, ACI Manual of Concrete Practices - Parts 1, 2, and 3, Detroit, MI.

American Petroleum Institute (API), 1973, Guide for Inspection of Refinery Equipment, Chapter II - Conditions Causing Deterioration or Failures, Second Edition, American Petroleum Institute Publishers, Washington, D.C.

_________, 1976, Guide for Inspection of Refinery Equipment, Chapter VI - Pressure Vessels, Third Edition, American Petroleum Institute Publishers, Washington, D.C.

_________, 1975, Guide for Inspection of Refinery Equipment - Chapter XII - Foundations, Structures and Buildings, Second Edition, American Petroleum Institute Publishers, Washington, D.C.

_________, 1981, Guide for Inspection of Refinery Equipment - Chapter XIII - Stmospheric and Low-Pressure Storage Tanks, American Petroleum Institute Publishers, Washington, D.C.

_________, 1974, Guide for Inspection of Refinery Equipment - Chapter XVI - Pressure Relieving Devices, Second Edition, American Petroleum Institute Publishers, Washington, D.C.

_________, 1970, Guide for Follow-up Inspection of Interior Tank Coatings, American Petroleum Institute Publishers, Washington, D.C.

_________, 1976, Cleaning Petroleum Storage Tanks, API Publication 2015, Second Edition, American Petroleum Institute Publishers, Washington, D.C.

_________, 1981, Recommended Practice of Abandonment or Removal of Used Underground Service Station Tanks, API Bulletin 1604, American Petroleum Institute Publishers, Washington, D.C.

American Society of Mechanical Engineers, 1971, Secion V - Nondestructive Examination ASME Boiler and Pressure Vessel Code, McGraw-Hill Publications, New York, NY.

American Water Works Association, 1953, Inspecting and Repairing Steel Water Tanks, Standpipes, Resevoirs and Elevated Tanks for Water Storage, AWWA Standard D101, American Water Works Association, Denver, CO.

Arney, H., 1978, Safe Entry and Cleaning of Storage Tanks and Vessels, National Petroleum Refiners Association, Technical Paper MC-7804.

British Valve Manufacturers Association, 1966, Valves For The Control of Fluids, Pergamon Press, Oxford.

Fitzgerald, J.H., "Corrosion Control for Buried Service Station Tanks," Paper No. 52, The International Corrosion Forum Devoted Exclusively to the Protection and Performance of Materials, April 14-18, 1975, Toronto, Canada, National Association of Corrosion Engineers, Publications Department, 1440 South Creek, Houston, TX, 77084.

Mahhotra V., 1976, Testing Hardened Concrete: Nondestructive Methods, American Concrete Institute, Detroit, MI.

Maryland Department of Natural Resources, Water Resources Administration, 1981, Maryland State Spill Contingency Plan, Fourth Edition, Annapolis, MD.

National Association of Corrosion Engineers (NACE), 1980, Corrosion Engineers Reference Book, National Association of Corrosion Engineers, Houston, TX.

National Fire Protection Association (NFPA), Water Resources Administration, 1981, Underground Leakage of Flammable and Combustible Liquids, National Fire Code, Section 329, Volume 13, National Fire Protection Association, Boston, MA.

_________, 1975, Cleaning Small Tanks and Containers, National Fire Code, Section 327, Volume 3, National Fire Protection Association, Boston, MA.

_________, 1981, Farm Storage of Flammable and Combustible Liquids, National Fire Code, Section 395, Volume 3, National Fire Protection Association, Boston, MA.

_________, 1981, Flammable and Combustible Liquids Code, National Fire Code, Section 30, Volume 3, National Fire Protection Association, Boston, MA.

_________, 1981, Pesticides in Portable Containers, National Fire Code, Section 43D, Volume 3, National Fire Protection Association, Boston, MA.

Perry, R., and C. Chilton, 1973, Chemical Engineer's Handbook, McGraw-Hill Publications, New York, NY

6. Personnel Health, Safety, and Training

6.1 INTRODUCTION

Two basic types of hazards to personnel exist in the workplace: safety hazards and health hazards. Safety hazards are those which can cause bodily injury, such as through a fall, an electric shock, or a cut. Health hazards are those which can cause illness or biologic damage, such as through exposure to toxic materials, including carcinogens or physical agents.

These hazards can be controlled by one or more of the following means:

- Elimination of the hazard by substituting a less hazardous material or process, by isolating the hazard so that workers are not exposed to it, or by removing the hazard altogether;

- Engineering control of the hazard by guarding equipment, by using mechanical means to reduce contaminant exposures to permissible limits, by altering equipment design or operation, or by using less hazardous methods or processes;

- Administrative control of the hazard by arranging work schedules or process times to reduce exposure, by training employees in safe work practices, and by administrative procedures and oversight designed to identify and abate hazards; or

- Use of personal protective equipment in cases where control of a hazard is not feasible or adequate by any of the first three methods, in emergency situations, or as a normal precautionary measure.

By the very nature of hazardous substance storage, hazard elimination will not be a realistic form of control. Engineering control, for example, by means of the technical guidelines provided by this manual, can be effective in reducing the risks associated with storage of hazardous chemicals. However, the degree of effectiveness is dependent upon rigorous implementation of administrative and personal protective controls. These include the proper selection and application of personal protective equipment, observance of safety precautions, and thorough training.

Standards and guidelines for minimizing hazards to personnel are addressed in Occupational Safety and Health Act (OSHA) regulations 29 CFR 1910 and 29 CFR 1926. The OSHA standards provide enforceable regulations intended to insure safe work environments. It is extremely important that the user of this manual determine the existence of and comply with state and local regulations which may supersede OSHA regulations. The standards are based on research from a variety of organizations. A sample listing of organizations which develop recommendations relevant to health and safety in hazardous materials environments includes: the National Institute for Occupational Safety and Health (NIOSH), the American National Standards Institute (ANSI), and the American Council of Governmental and Industrial Hygienists (ACGIH). These organizations publish recommendations covering a wide variety of health and safety issues, which generally are more frequently updated than the OSHA standards. Therefore, although they are not enforceable unless specifically cited by OSHA, the recommendations of these organizations should be used to supplement the OSHA standards in order to establish the most effective health and safety practices.

6.2 HEALTH HAZARDS

Health hazards result from exposure to certain contaminants or physical stresses that can cause illness or biologic damage. They may be chemical or physical in nature. Benzene, for example, is a chemical hazard, whereas excessive noise exposure is a physical hazard.

At present, the OSHA standards contain regulations governing employee exposure to chemical and physical hazards. Exposure to these hazards can have a variety of adverse effects on the human system. These include, but are not limited to, skin disease, respiratory damage, and sensory damage.

Factors that determine the degree of hazard are the concentration of the contaminant in the environment, and the length of time workers are exposed to the contaminant. Permissible exposure limits and standards can be found in the OSHA regulations (29 CFR 1910), NIOSH criteria documents, and the ACGIH Threshold Limit Values booklet.

The major health hazards associated with hazardous substance storage systems are those resulting from skin and respiratory exposure. These hazards, their causes, the best methods of control, and applicable control standards are discussed in the following sections.

6.2.1 Skin Disease

Occupational dermatitis is the most frequently encountered job-related disease. One out of every four workers is exposed to some form of skin irritant, and about one percent of those exposed develop skin disorders. Occupational dermatitis may be caused by primary irritants, allergic sensitizers, mechanical trauma, plant poisons, and biologic agents.

- Primary irritants affect anyone who comes into direct skin contact with them. They produce skin irritation at the point of contact. Many solvents, lubricants, acids, and caustics are common primary irritants.

- Allergic sensitizers can, after a period of time, cause an allergic-type skin irritation in susceptible people. Typical sensitizers include epoxy resin hardeners and coal tar derivatives.

- Mechanical trauma is skin irritation resulting from friction, pressure, or other mechanical means.

- Plant toxins include poison ivy and poison oak, which produce irritations of the skin.

- Biologic agents include those bacteria, fungi, and parasites which attack the skin and produce irritation.

Skin cancer is in a category by itself. It may be occupationally caused by worker contact with known or suspected carcinogenic agents, or it may be caused by excessive exposure to ultraviolet or ionizing radiation.

The OSHA standards contain no sections devoted specifically to skin disease. The general standards for workplace sanitation in Subpart J of 29 CFR 1910 are primarily concerned with control of infectious diseases, but may be of benefit in reducing skin exposures. Further guidance may be obtained from the permissible concentrations for various workplace contaminants listed in Tables Z-1, Z-2, and Z-3 of the Air Contaminants section in Subpart Z. Those substances followed by the word "skin" can be absorbed through the skin. Standards for many specific toxic and hazardous substances are addressed in Subpart Z, as well as in the ACGIH Threshold Limit Values booklet, and the NIOSH Pocket Guide to Chemical Hazards.

Local exhaust ventilation is a good method for controlling most air contaminants, including those aerosols that are harmful to the skin. Further guidance on ventilation may be obtained from ANSI standards Z9.1 and Z9.2, and from the ACGIH Industrial Ventilation Manual. However, the best methods for controlling direct skin exposure are through the use of protective clothing, as described in Section 6.3.1 of this manual.

6.2.2 Respiratory Damage and Disease

Permissible concentrations listed by NIOSH, ACGIH, and OSHA generally are levels of contaminants to which nearly all workers may be exposed day after day without developing significant adverse effects. If the permissible concentrations for air contaminants are regularly exceeded, workers may experience such effects as respiratory damage or disease, systemic disorders, chronic or irreversible tissue changes, or narcosis of sufficient degree to impair physical ability.

Air contaminants that affect health may be either particulate or gaseous in nature. Dusts and fumes are particulates. Dusts are tiny solid particles suspended in the atmosphere. Fumes are tiny particles resulting from condensation of volatilzed metal, such as iron or lead. Gaseous contaminants may be true gases (such as carbon monoxide) or they may be vapors of substances that are liquid at normal temperatures.

The effect caused by particulate and gaseous contaminants may be acute or chronic. An acute effect occurs when the concentration of contaminant is so great that some system within the body is completely overwhelmed and can no longer perform its vital function. A chronic effect is one that results from exposure to lower contaminant concentrations over long time periods. Such exposure could ultimately lead to disability or death.

The permissible concentrations for air contaminants should be strictly adhered to. Exposure of employees to contaminant concentrations greater than those allowed by the standards may be harmful to employee health. Therefore, the workplace air should be monitored regularly to assure that contaminant concentrations are well within the permissible limits. Workplace monitoring for air contaminants should be done by trained, qualified personnel who know how to use the measuring instruments and how to interpret the values obtained from sampling the air. The air monitoring program should be developed and supervised by a Certified Industrial Hygienist. Because sampling protocols vary by the type of hazard to be measured and by the objectives of the sampling program (i.e., personal exposure, ambient levels, etc.), specific guidance should be obtained from such sources as:

- NIOSH's <u>Pocket Guide to Chemical Hazards</u>;
- NIOSH's <u>The Industrial Environment-Its Evaluation and Control</u>; and
- Patty's <u>Industrial Hygiene and Toxicology</u>.

When deciding on abatement measures, each type of exposure must be considered separately. Some controls, as stated in the standards, are mandatory, while others must be adapted to the situation at hand. Any abatement measures used, including the use of personal protective equipment, must be approved by an industrial hygienist or comparably qualified person.

The following OSHA standards (29 CFR 1910) are applicable to control of respiratory hazards at hazardous materials storage facilities:

- Subpart I covers respirators for particular hazards, including regulations for adequate fit, maintenance, and proper use;
- Subpart Q provides regulations for respiratory protection during welding, cutting, and brazing operations; and
- Subpart Z specifies the permissible concentrations for a wide range of air contaminants. It also explains how exposures must be calculated and the kinds of abatement measures that must be undertaken.

Other standards which should be considered and which may provide applicable guidance, though not regulatory in nature, include:

- ANSI Z88.2-Practices for Respiratory Protection;
- ANSI Z9.1-Safety Code for Ventilation and Operation of Open Surface Tanks;
- ANSI Z9.2-Fundamentals Governing the Design and Operation of Local Exhaust Systems;
- NIOSH Pocket Guide to Chemical Hazards: And
- ACGIH Threshold Limit Values for Chemical Substances and Physical Agents in the Workroom Environment.

6.3 PERSONAL PROTECTIVE EQUIPMENT

Personal protective equipment must be worn by hazardous substance storage facility personnel to prevent injuries from a variety of hazards, such as contact with hot or corrosive liquids, inhalation of toxic gases or fumes, heavy materials or equipment accidents, or injuries with objects. Protection against electrical shock must be provided if conditions warrant. Additional safety devices are required for workers in high places or over water.

The general regulations for personal protective equipment are found in 29 CFR 1910.132. These requirements state that protective equipment, respiratory devices, and protective shields and barriers, shall be provided, used, and maintained in a sanitary and reliable condition wherever necessary because of physical, chemical, or mechanical hazards capable of causing injury or impairment in the function of any part of the body through absorption, inhalation, or physical contact. All personal protective equipment shall be of a design and construction appropriate to the work to be performed.

In some situations which require the use of personal protective equipment, workers may resist using the equipment because it is uncomfortable or viewed as an inconvenience. Supervisors are responsible for ensuring that workers use the prescribed equipment in an appropriate manner.

Procedures that may help to gain worker acceptance of personal protective equipment include:

- An educational program pointing out the necessity for and benefits of such equipment;
- Appropriate rules to influence worker acceptance; and
- Selection of equipment that provides adequate protection with minimal interference with normal work procedures.

It is also the supervisor's responsibility to make sure that all personal protective equipment is periodically checked and properly maintained.

Because the major health and safety hazards associated with hazardous substances are due to skin and respiratory exposure, standards for protection in these areas will be discussed in the sections to follow. The user is referred to the appropriate OSHA (20 CFR 1910.133, .135, and .136) and ANSI (Z41.1, Z87.1, and Z89.1) standards for face, head, and foot protection.

6.3.1 Body and Hand Protection

At present there are no specific OSHA or ANSI standards for body and hand protection. However, the general requirements of OSHA Standard 1910.132 state that personal protective equipment must be provided where conditions warrant.

Many types of specialized clothing are available to provide protection against a variety of hazards. In general, the design, construction, and material used should provide appropriate protection for the hazard involved. The manufacturer often provides guidelines regarding appropriate usage.

Materials used for body and hand protection differ, depending on the type of protection needed.

- Leather protects against heat, hot metal splashes, and infrared and ultraviolet radiant energy;
- Asbestos and wool are used for heat protection at higher temperatures;
- Aluminized clothing is used at extremely high temperatures to reflect much of the radiant heat;
- Padded clothing and hard fiber or metal shields protect against bruises, cuts, and blows.
- Impervious clothing is required for protection against dusts, vapors, moisture, and corrosive liquids. Such garments run the gamut from sheet plastic bibs to total body suits with an air supply.

Impervious materials include natural rubber, synthetic rubber, neoprene, vinyl, polypropylene, and polyethylene film. Natural rubber is not suitable for use with oils, greases, and many organic solvents and chemicals. Table 6-1 lists a selection of impervious materials and indicates the level of protection they provide against a variety of chemicals. The data in the table are based solely on manufacturers' information, with no guarantee of their accuracy of reliability. It is recommended that the manufacturers be contacted for verification before selecting materials for specific applications. Factors to be considered include: chemical composition of the materials to be encountered, the degree of concentration, temperature conditions,

Table 6-1

CHEMICAL RESISTANCE OF PROTECTIVE CLOTHING MATERIALS

	Neo-prene	PVC	Paracril/ PVC	Polyur-ethane	Chlorinated Polyethy-lene	Butyl Rubber	Natural Rubber	Nitrile	Vitron	PVA
Acetaldehyde	C	C	A	C	I	A	A	C	U	U
Acetic Acid	A	C	A	C	A	C	A	A	U	U
Acetone	A	C	U	C	A	C	A	C	U	U
Acrylonitrile	A	C	A	A	A	C	C	C	I	I
Ammonium Hydroxide	A	A	A	A	A	A	A	A	A	U
Amyl Acetate	C	C	A	C	C	C	C	C	U	A
Aniline	C	C	A	A	A	C	C	C	C	C
Benzaldehyde	C	C	A	C	C	A	C	C	U	A
Benzene	C	U	C	A	C	U	U	C	C	A
Benzyl Alcohol	A	C	A	C	A	A	A	A	I	I
Benzyl Chloride	C	C	A	C	U	C	C	C	C	I
Butyl Acetate	C	C	A	C	A	C	C	C	U	A
Butyl Alcohol	A	A	A	C	A	A	A	A	I	C
Carbolic Acid	C	C	A	A	A	C	A	C	A	C
Carbon Disulfide	C	C	A	C	C	U	C	C	I	A
Carbon Tetrachloride	C	C	A	C	C	U	C	A	A	A
Chloroacetone	C	U	C	C	U	C	C	C	I	I
Chloroform	C	C	A	C	U	U	C	C	A	A

A=Acceptable
U=Unacceptable
C=Conditionally Acceptable
I=Insufficient Data.

Note: This table is provided as a guide only. The user is advised to contact the protective clothing manufacturer regarding the specific applicability and limitations of a material under proposed conditions of use.

Table 6-1 (Cont.)

	Neo-prene	PVC	Paracril/ PVC	Polyur-ethane	Chlorinated Polyethy-lene	Butyl Rubber	Natural Rubber	Nitrile	Vitron	PVA
Coal Tar Products	C	C	A	A	C	C	C	C	I	I
Cyclohexane	C	C	U	A	A	U	C	A	A	I
Diacetone Alcohol	A	A	A	C	A	A	A	A	I	I
Dibutyl Phthalate	C	C	A	C	C	A	C	A	A	A
Ethanol	A	A	A	C	I	A	A	A	A	U
Ethyl Ether	C	C	A	C	A	C	C	A	U	A
Ethylene Glycol	A	A	A	C	A	A	A	A	A	C
Formaldehyde	C	A	A	C	A	C	A	A	A	U
Formic Acid	A	A	A	A	A	A	A	C	U	U
Furfural	A	C	A	C	A	A	A	C	U	C
Gasoline	A	C	A	A	A	U	C	A	A	A
Glycerine	A	A	A	C	A	A	A	A	A	A
Hydrobromic Acid	A	A	A	C	A	A	I	I	I	I
Hydrochloric Acid	A	A	A	U	A	A	A	A	A	U
Hydrofluoric Acid	C	C	C	U	A	A	A	A	A	U
Hydrogen Peroxide	A	A	A	A	A	C	A	A	C	U
Hydrogen Sulfide	A	A	A	U	A	A	I	I	A	I
Isopropyl Alcohol	A	A	A	C	A	A	A	A	A	U
Kerosene	A	C	A	A	A	U	C	A	A	A
Lactic Acid	A	A	A	A	A	A	A	A	A	C
Linseed Oil	A	A	A	C	A	C	C	A	A	A

A=Acceptable
U=Unacceptable
C=Conditionally Acceptable
I=Insufficient Data

Note: This table is provided as a guide only. The user is advised to contact the protective clothing manufacturer regarding the specific applicability and limitations of a material under proposed conditions of use.

Table 6-1 (Cont.)

	Neo-prene	PVC	Paracril/PVC	Polyur-ethane	Chlorinated Polyethy-lene	Butyl Rubber	Natural Rubber	Nitrile	Vitron	PVA
Malic Acid	A	A	A	C	I	U	I	I	I	I
Methyl Acetate	A	C	C	C	A	A	C	C	I	I
Methanol	A	A	A	C	A	A	A	A	C	U
Methyl Ethyl Ketone	C	U	C	C	C	A	A	U	U	C
Nitric Acid	C	C	A	U	C	C	C	C	A	U
Nitrobenzene	C	C	A	C	C	U	C	C	A	I
Oleic Acid	A	A	A	C	A	A	C	A	A	A
Perchloroethylene	C	U	A	C	C	U	U	A	A	A
Phosphoric Acid	A	A	A	C	A	A	A	A	A	U
Pine Oil	A	A	A	C	A	U	C	A	I	I
Potassium Hydroxide	A	A	A	C	A	A	A	A	C	U
Sodium Hydroxide	A	C	A	C	A	C	A	A	C	U
Sulfuric Acid	C	C	A	U	C	C	C	C	C	U
Tannic Acid	A	A	A	A	A	A	A	A	A	U
Toluene	C	U	A	A	C	U	C	C	A	A
Trichloroethylene	C	C	A	C	C	U	C	C	A	A
Triethanolamine	A	A	A	C	A	A	A	A	U	A
Turpentine	C	C	A	C	A	U	C	A	A	A

A=Acceptable
U=Unacceptable
C=Conditionally Acceptable
I=Insufficient Data

Source: Ecology and Environment, Inc., 1982, from manufacturers' data.

Note: This table is provided as a guide only. The user is advised to contact the protective clothing manufacturer regarding the specific applicability and limitations of a material under proposed conditions of use.

abrasive effects of the materials being handled, and the length of time users will be in contact with the material.

Gloves, hand leathers, and arm protectors are available for protection against heat, chemicals, abrasions, and slippery surfaces. Hand leathers may be more comfortable than gloves for heavy materials handling but should not be used around moving machinery. Gloves should be snug but comfortable and extend above the wrist so that no gap is left between glove and coat or shirt sleeve.

Clothing available for use around specific hazards includes the following:

- High visibility clothing for use where visibility is important;
- Disposable clothing for use where chemical contamination is a problem; and
- Non-conductive clothing for use during bare-hand work on high voltage conductors.

Protective clothing must be of good quality and well-constructed. Clothing fasteners must prevent gaps during body movement, and must be designed so that the wearer can remove the garment rapidly and easily. Clothing used around hot liquids, molten metals, acids, and caustics must not have turned-up cuffs or other projections. Pockets should have flaps that fasten shut.

6.3.2 Respiratory Protection

According to OSHA standards, the primary means of controlling respiratory hazards shall be the prevention of atmospheric contamination. This shall be accomplished, as far as feasible, by accepted engineering control measures, such as enclosure or confinement of the operation, general and local ventilation, and use of the least toxic materials. When effective engineering controls are not feasible, or while they are being instituted, appropriate respiratory protection must be used.

According to OSHA General Requirements for Respiratory Protection (29 CFR 1910.134), the employer is responsible for establishing and maintaining a respiratory protective program. Elements of a minimally acceptable program include the following:

- Written standard operating procedures governing the selection and use of respirators shall be established;
- Respirators shall be selected on the basis of hazards to which the worker is exposed.
- The user shall be instructed and trained in the proper use of respirators and their limitations.

- Where practical, respirators should be assigned to individual workers for their exclusive use.
- Respirators shall be regularly cleaned and disinfected.
- Respirators shall be stored in a convenient, clean, and sanitary location.
- Respirators used routinely shall be inspected during cleaning, and defective parts replaced. Those for emergency use should be thoroughly inspected after each use, and at least once a month.
- Appropriate surveillance of work area conditions and appropriate records on degree of employee exposure or stress shall be maintained.
- There shall be a regular inspection and evaluation to determine the program's continued effectiveness.
- Persons should not be assigned to tasks requiring the use of respirators unless it has been determined that they are physically able to use the equipment and perform the work.
- Approved or accepted respirators should be used when they are available.

Selection of respirators should be made in accordance with the standards in ANSI Z88.2. Proper respirator selection will require consideration of the following factors:

- Nature of the hazard;
- Characteristics of the hazardous operation or process;
- Location of the hazardous area with respect to safe breathing areas;
- Time period for which respiratory protection may be needed;
- Activity of the workers in the hazardous area;
- Physical characteristics, capabilities, and limitations of various respirator types; and
- Respirator fit and protection factors.

Table 6-2 provides guidelines for respirator selection on the basis of the hazard to be encountered. Detailed selection criteria are found in ANSI Z88.2-1980. Additional guidance for permissible exposure limits, air monitoring procedures, and appropriate respirator types is found in the NIOSH Pocket Guide.

Air-supplied respirators are required in oxygen-deficient (less than 19.5% oxygen) atmospheres, and may also be used where

Table 6-2

SELECTION OF RESPIRATORS

Hazard	Respirator (see note)
Oygen deficiency, immediately dangerous to life and health.	Air-line, continuous-flow, pressure-demand type with escape provisions. Air-line, continuous-flow helmet, hood, or suit, with escape provisions. Self-contained breathing apparatus (pressure-demand type, or positive-pressure, closed-circuit type).
Oxygen deficiency, not immediately dangerous to life and health.	Self-contained breathing apparatus. Hose mask with blower. Combination air-line respirator with auxiliary self-contained air supply.
Gas and vapor contaminants immediately dangerous to life and health.	Self-contained breathing apparatus (pressure-demand-type, open-circuit, or positive-pressure closed circuit). Powered air-purifying, full facepiece respirator with chemical cannister (if escape provisions are provided). Self-rescue mouthpiece respirator (for escape only). Combination air-line respirator with auxiliary self-contained air supply.
Gas and vapor contaminants not immediately dangerous to life and health.	Air-line respirator. Hose mask with or without blower. Air-purifying respirator with chemical cartridge.
Particulate contaminants immediately dangerous to life and health.	Self-contained breathing apparatus (pressure-demand-type open-circuit, or positive-pressure, closed-circuit). Air-purifying, full facepiece respirator with appropriate filter (if escape provisions are provided). Combination air-line respirator with auxiliary self-contained air supply.
Particulate contaminants not immediately dangerous to life and health.	Air-purifying, respirator with particulate filter pad or cartridge. Air-line respirator. Air-line, continuous flow helmet, hood or suit. Hose mask with or without blower.
Combination gas, vapor, and particulate contaminants immediately dangerous to life and health.	Self-contained breathing apparatus (pressure-demand-type open-circuit, or positive-pressure, closed-circuit). Air-purifying, full-facepiece respirator with chemical cannister and appropriate filter (if escape provisions are provided). Combination air-line respirator with auxiliary self-contained air supply.
Combination gas, vapor, and particulate contaminants not immediately dangerous to life and health.	Air-line respirator. Hose mask with or without blower. Air-purifying respirator with chemical cartridge and appropriate filter.

Note: For the purpose of this table, "immediately dangerous to life and health" is defined as any atmosphere that poses an immediate hazard to life, or produces immediate, irreversible debilitating effects on health. Consult ANSI Z88.2-1980 for further definition and clarification of respirator selection criteria.

Source: 29 CFR 1926.103 and ANSI Z88.2 - 1980.

concentrations of toxic gases or vapors exceed the permissible concentration. Such respirators include self-contained breathing apparatuses (SCBA) or air-line respirators which supply air to individuals from a central source. If the atmosphere is determined to be immediately dangerous to life and health (IDLH), the air supply must be of the continuous-flow type, with escape provisions, or it may be a pressure-demand or positive-pressure, closed-circuit SCBA (see Table 6-2 and ANSI Z88.2-1980).

Air-purifying respirators are used to remove gaseous and particulate contaminants. They do not protect against oxygen deficiency, and must only be used in atmospheres with more than 19.5% oxygen. Further limitations of air-purifying respirators require that they be used only to protect against known contaminants with adequate warning properties; that the appropriate filters, cartridges, or cannisters be in place; that the user be appropriately fit-tested; that contaminant levels be continously monitored; and that the contaminant levels do not exceed the capabilities of the respirator. Air-purifying respirators generally are not acceptable for use in IDLH atmospheres, unless they are of the powered type and have appropriate escape provisions (see Table 6-2 and ANSI Z88.2-1980). Types of air-purifying respirators include particulate-removing, vapor- and gas-removing, and combination particulate- and vapor- and gas-removing respirators.

Particulate removing respirators are used to remove low concentrations of particles by drawing air through a filter. As the contaminant is deposited on the filter, the filtration efficiency increases, but resistance to air flow also increases. In higher particulate concentrations, the filters may become clogged and make breathing difficult.

Vapor- and gas-removing respirators consist of a half- or full-mask facepiece equipped with cartridge(s) or cannister(s) to remove a single vapor or gas, a single class of vapors or gases, or a combination of two or more classes of vapors or gases. These cartridges and cannisters are filled with activated charcoal or some other material that will adsorb gases and vapors. The respirators lose effectiveness when the sorbents become saturated. Because no single sorbent will remove all types of gaseous contaminants, the cannister must be chosen to fit the specific need. Vapor- and gas-removing cartridges and cannisters are often combined with particulate-removing filters to provide additional protection. Auxiliary self-contained supplies of respirable air can be used in addition to an air-purifying respirator to provide means of escape in the event of emergency.

For safe use of any respirator, the user must be properly trained in its selection, use, and maintenance. Every respirator wearer shall receive instructions in how to wear the respirator and how to determine if it fits properly. Respirators shall not be worn when conditions prevent a good face seal. Such conditions include:

- Facial hair;
- Protruding skull cap;
- Eyeglass temple pieces; and
- Absence of one or both dentures.

Respirator fit may be determined by exposing the wearer to an irritant smoke or odorous vapor. The wearer is then instructed to perform a series of movements to simulate work activity, and an observation is made to determine if the irritant or odor is detected. The fit test may be performed qualitatively or quantitatively. In the qualitative test, the adequacy of fit is determined by the wearer's detection of (or reaction to) test smoke or vapor. In a quantitative test, instrumentation is used to measure the test atmosphere and the air inside the respirator to determine the extent of penetration of the test agent into the respirator. Records should be maintained of fit test results. Under optimal conditions, with an acceptable facepiece fit, protection factors such as those in Table 6-3 can be achieved. The protection factor is the ratio of concentration of contaminant in the ambient atmosphere to that inside the facepiece under conditions of use.

In order to insure continued effectiveness of the respirators, they must be regularly inspected, cleaned, disinfected, repaired, and properly stored. Inspection should include checking tightness of connections and the condition of the facepiece, headbands, valves, connecting tube, and cannisters. Records should be kept of inspection dates and findings. Routinely used respirators should be cleaned and disinfected after each use in accordance with the manufacturer's directions. Any necessary repairs should be performed only by experienced persons using parts designed for the respirator. After inspection, cleaning, and any necessary repair, the respirator should be stored so that the facepiece and exhalation valve will rest in a normal position, and should be protected against dust, sunlight, heat, extreme cold, excessive moisture, or damaging chemicals. Respirator users should always refer to the manufacturer's instructions and ANSI Z88.2 for specific guidance on operation and maintenance procedures, including respirator testing, air supply selection and testing, and cleaning.

6.4 ACTIVITIES IN HAZARDOUS AREAS

Work in areas of flammable liquids and vapors and in confined spaces requires additional caution on the part of the employee. In those where ignitable vapors may be expected, no metal cutting, brazing, or welding should be performed unless the area is monitored and found to be free of ignitable vapor concentrations. Care should be taken in these areas to see that metal equipment, such as tanks and transfer pipes, are adequately grounded and bonded before operating the equipment or transferring any flammable materials. Non-sparking tools and intrinsically safe electrical equipment should be used whenever possible. Smoking should be prohibited except in designated safe areas.

Work in confined spaces poses risks of inadequate ventilation or oxygen supply, accumulation of toxic or combustible materials, or entrapment. Confined spaces generally have limited means of access, or are so enclosed that inadequate dilution ventilation can occur. Examples of confined spaces include, but are not limited to: storage

Table 6-3

OPTIMAL RESPIRATORY PROTECTION FACTORS

Type Respirator	Protection Factor
I. Air purifying:	
A. Particulate removing	
Single-use, dust	5
Quarter mask, dust	5
Half mask, dust	10
Half or quarter mask, high efficiency	10
Half or quarter mask, fume	10
Full facepiece, high efficiency	50
Powered, high efficiency, all enclosures	1,000
Powered, dust or fume, all enclosures	
B. Gas and vapor removing	10
Half mask	50
Full facepiece	
II. Atmosphere supplying:	
A. Supplied air	
Demand, half mask	10
Demand, full facepiece	50
Hose mask without blower, full facepiece	50
Pressure demand, half mask	1,000
Pressure demand, full facepiece	2,000
Hose mask with blower, full facepiece	50
Continuous flow, half mask	1,000
Continuous flow, full facepiece	2,000
Continuous flow, hood, helmet, or suit	2,000
B. Self contained breathing apparatus (SCBA)	
Open circuit, demand, full facepiece	50
Open circuit, pressure demand full facepiece	10,000
Closed circuit, oxygen tank-type, full facepiece	50

Source: Clayton and Clayton, 1978.

tanks, tank cars, bins, silos, manholes, tunnels, pipelines, ovens, and other similar structures.

Before entry into confined spaces, all lines containing harmful agents entering the space should be physically disconnected or blocked. Any fixed mechanical or electrical equipment should be disconnected. The atmosphere within the space should be tested for oxygen level, explosive gas, and if there is reason to suspect their presence, for air contaminants in excess of permissible exposure limits. If the tests indicate the area is unsafe (having less than 19.5% oxygen, greater than 10% lower explosive limit of an explosive gas, or an excessive contaminant level), the area must be ventilated until safe levels are achieved. If safe levels are not achieved, entry may be made only by using a self-contained or supplied-air breathing apparatus. A person may not remain in the confined space when the primary air supply is depleted or being replaced. Moreover, no one should enter the confined space unless there is provision for constant audible and visual communication, and adequate rescue procedures have been outlined. Employees working both inside and outside the confined space should be adequately trained in rescue, first aid, and cardio-pulmonary resuscitation procedures. It is recommended that a continuous supply of air be provided to the confined space while work is being performed, particularly if toxic or combustible gases were originaly present, if organic solvents are being used, or if open-flame torches are in use. In the event of an emergency, persons attempting rescue must wear appropriate respiratory protection.

For work in hazardous areas, special attention must be paid to the development and documentation of emergency contingency planning. These plans should address such details as levels of personal protection required, identification of emergency resources, and delineation of safety or evacuation zones.

Levels of protection and safety zones are determined by appropriate air monitoring data and expected air dispersion patterns. These patterns may be calculated by methods such as described in Section 3.7, and will provide indications of expected contaminant concentrations at various distances from the site. In lieu of air dispersion modeling, the zones may be determined by determining wind direction and speed and safety zones, based on such sources as the DOT Emergency Action Guide. In all cases, access to the site should be from the upwind side, with appropriate air monitoring equipment, such as oxygen detectors, explosive vapor detectors, and toxic gas detectors. In general, atmospheres containing less than 19.5% oxygen by volume at sea level are considered to be "oxygen deficient" and require use of an air supply. Explosive gas levels measured in excess of 10% of the lower flammability limit (LFL) dictate caution; in excess of 40%, evacuation is recommended. (Note: Action levels based upon the measured percentage of LFL may be varied under specific circumstances, upon consideration of such factors as the known nature of the hazard, emergency preparedness, extent of threat to life and property, nature of tasks to be performed within the hazardous area, and applicable regulations). Selection of the appropriate respirator can be made in accordance with guidelines such as are given in Table 6-2, the NIOSH Pocket Guide, and ANSI Z88.2-1980.

For persons working in contact with hazardous materials, appropriate contamination reduction measures must be employed. These include use of disposable protective clothing and thorough decontamination before leaving the site. Decontamination areas should be located upwind of the facility, and access should be limited to persons entering the site and those assisting with decontamination.

In general, decontamination consists of rinsing personnel and equipment with copious amounts of water and washing with a detergent solution. If specific contaminants are known, then a specific detergent or solvent can be used as a decontaminant.

Decontamination solutions should be designed to react with and neutralize the contaminants which may be encountered. In many cases, it may be necessary to use a solution effective for a variety of contaminants. Several types of general purpose decontamination solutions which can be prepared from easily obtained materials are:

- Type A - A solution containing 5% sodium carbonate and 5% trisodium phosphate;
- Type B - A solution containing 10% calcium hypochlorite;
- Type C - A solution containing 5% trisodium phosphate (may also be used as a general purpose rinse); and
- Type D - A dilute solution of hydrochloric acid.

Recommended applications of these solutions are shown in Table 6-4.

In practice, decontamination is a stepwise process moving from the highest potential contamination area to the lowest. The sequence for decontamination in a "worst-case" situation is illustrated in Figure 6-1. In this case a "hot line" distinguishes an area assumed to be contaminated from a contamination reduction area (or exclusion zone), and another line distinguishes the contamination reduction zone from the safe (non-exclusion) area; initial washing and rinsing is performed on the contaminated side of the "hot line," and subsequent washing and equipment removal is performed in the contamination reduction (exclusion) area. No individuals should leave the site without undergoing thorough decontamination. This should be performed only by persons stationed in the contamination reduction area, and these persons should wear protective clothing and equipment appropriate to the hazards they may encounter during the decontamination process.

6.5 FIRST AID AND MEDICAL SURVEILLANCE

In the event of an accident at a hazardous materials facility, inadequate first aid facilities or lack of trained personnel may result in deaths and permanent disabilities. Improper rescue and transport of an injured person may result in further injury. Costs to the facility may be high in terms of medical care, insurance, and lost earnings.

Table 6-4

USES OF GENERAL PURPOSE DECONTAMINATION SOLUTIONS

Type of Hazard Suspected	Solution Type	Solution
1. Inorganic acids, metal processing wastes.	A	To 10 gallons of water, add four pounds of sodium carbonate (soda lime) and four pounds of trisodium phosphate. Stir until evenly mixed.
2. Heavy metals: mercury, lead, cadmium, etc.	A	To 10 gallons of water, add four pounds of sodium carbonate (soda lime) and four pounds of trisodium phosphate. Stir until evenly mixed.
3. Pesticides, fungicides, chlorinated phenols, dioxins, and PCBs.	B	To 10 gallons of water, add eight pounds of calcium hypochlorite. Stir with wooden or plastic stirrer until evenly mixed.
4. Cyanides, ammonia, and other non-acidic inorganic wastes.	B	To 10 gallons of water, add eight pounds of calcium hypochlorite. Stir with wooden or plastic stirrer until evenly mixed.
5. Solvents and organic compounds, such as trichloroethylene, chloroform, and toluene.	C (or A)	To 10 gallons of water, add four pounds of trisodium phosphate. Stir until evenly mixed.
6. PBBs and PCBs.	C (or A)	To 10 gallons of water, add four pounds of trisodium phosphate. Stir until evenly mixed.
7. Oily, greasy unspecified wastes.	C	To 10 gallons of water, add four pounds of trisodium phosphate. Stir until evenly mixed.
8. Inorganic bases, alkali, and caustic waste.	D	To 10 gallons of water, add one pint of concentrated hydrochloric acid. Sir with a wooden or plastic stirrer.

Source: Ecology and Environment, Inc., 1982.

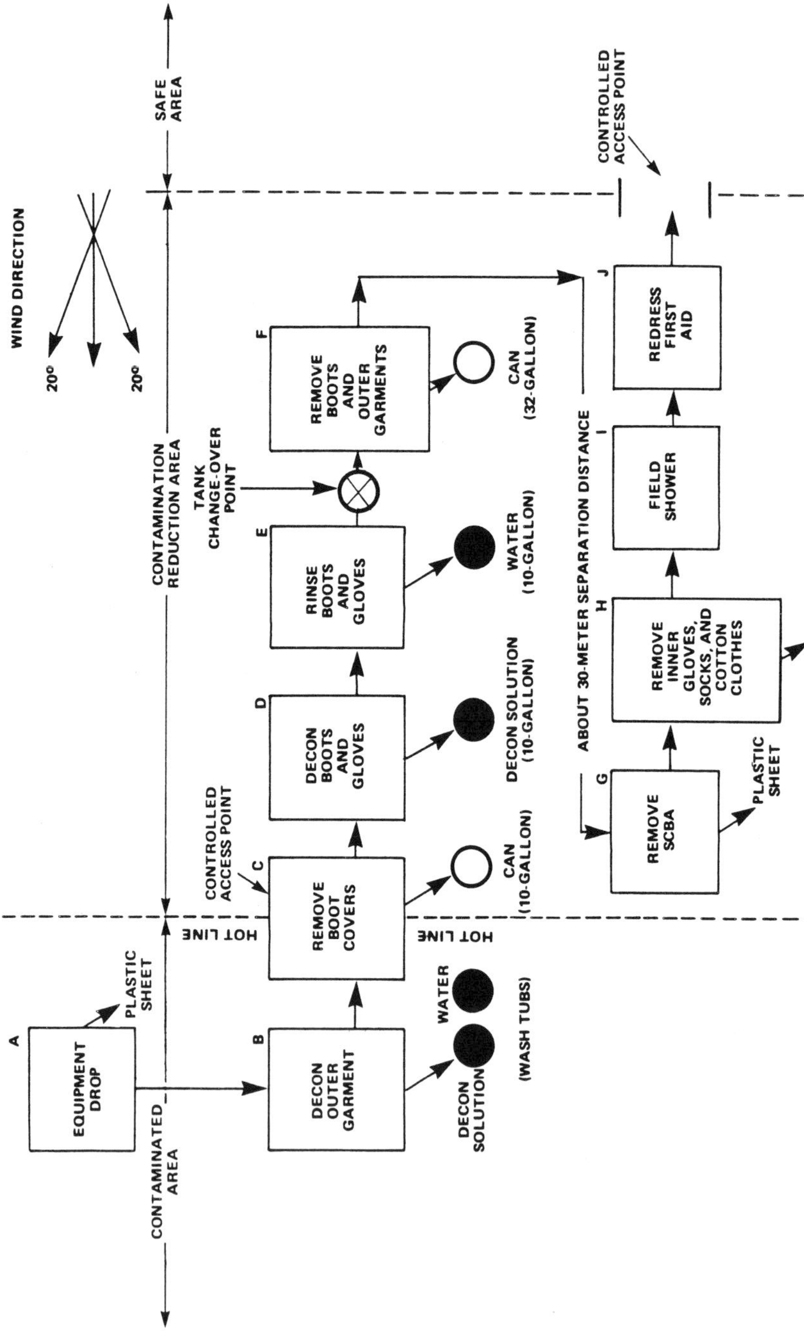

SOURCE: Ecology and Environment, Inc., 1983.

Figure 6–1 MAXIMUM LAYOUT OF PERSONNEL DECONTAMINATION STATION

The OSHA standards (29 CFR 1910.151) for medical and first aid services/facilities are brief but important. They are as follows:

- The employer must insure that medical personnel are readily available for advice and consultation;

- In the absence of an infirmary, clinic, or nearby hospital for treatment, at least one person trained in first aid must be available. All first aid facilities and supplies must be approved by the establishment's consulting physician;

- Where the eyes or body of any person may be exposed to injurious corrosive materials, suitable facilities must be provided in the work area for quick drenching or flushing of the eyes or body in emergencies.

Although the OSHA standards for the construction industry are not directly applicable to storage facilities, these regulations are worth noting as guidelines. The construction standards for medical and first aid are similar to those in 29 CFR 1910.151, but a bit more detailed. These regulations are as follows:

- The person(s) available at the work site to render first aid must have a valid first aid training certificate from the U.S. Bureau of Mines, the Red Cross, or the equivalent;

- First aid supplies approved by the consulting physician must be easily accessible. The first aid kit must be weatherproof with individual sealed packages for each type of item;

- The contents of the kit must be checked by the employer to make sure depleted items are replaced before the kit is taken out on any job;

- Proper equipment must be provided for transporting injured persons to a doctor or hospital. If this kind of transportation cannot be provided, then a communication system for contacting an ambulance must be established; and

- The telephone numbers of physicians, hospitals, and an ambulance must be conspicuously posted.

These standards constitute the basic elements of a first aid program. Such a program should address basic first aid procedures, but should also emphasize the particular hazards at each facility. It is appropriate that, as part of the first aid program, a safety plan be prepared for each maintenance job. The plan should outline such items as: specific first aid procedures for the hazards expected to be encountered; addresses, maps, and direction to the nearest emergency facilities; names and phone numbers of trained medical emergency response personnel; and pertinent medical information (allergies, etc.) about on-site personnel which may be of benefit to attending medical care providers.

The first aid program should be part of an overall medical surveillance program. Persons working routinely with hazardous materials should be provided with pre-employment physical examinations to determine their basic fitness for performing the job, as well as to provide baseline data on systemic functions which may be affected by job conditions. Depending upon the frequency and extent of hazardous materials exposure, follow-up examinations should be provided on at least an annual basis. In the event of an accidental exposure, immediate examinations are warranted. The medical surveillance program should be directed by a physician specializing in occupational medicine.

The day-to-day administration of the first aid and medical surveillance program may be charged to a Facility Safety Coordinator. This person should be an industrial hygienist or someone with equivalent training and experience. This person would have such responsibilities as:

- Maintaining an on-going first aid and safety procedures training program;
- Providing technical expertise in health and safety matters;
- Overseeing the facility's compliance with health and safety standards;
- Preparing and/or reviewing job-related safety plans; and
- Serving as the point of reference for all health and safety conflicts at the facility.

6.6 TRAINING

Experienced, well trained people are essential for successful implementation of a containment assurance and safety program. Training sessions must be held on a regular basis because duties of maintenance and inspection personnel may change, new equipment may be acquired, and containment assurance techniques may be modified as more experience is gained. The supervisors of inspection, maintenance, and emergency personnel, under the direction of the Facility Safety Coordinator, should be responsible for providing the continued training.

The objective of the containment assurance and safety program is to reduce the likelihood of an accidental release of hazardous substances, and, in the event of a spill, to effect a safe, quick, and efficient clean-up with minimal adverse effect on people or the environment. The purpose of the training program is to prepare storage site personnel to achieve this objective. The training program should consist of monthly or quarterly meetings which can include classroom exercises, field training, and response drills. The specific goals of the training program are to teach the response team members to:

- Use monitoring and protective equipment correctly;
- Identify potential trouble spots;
- Implement the containment assurance and safety program; and
- Critically review and upgrade the program.

Most of the training should be oriented toward drills and demonstrations rather than formal classroom instruction. However, certain aspects of the training can be covered efficiently in group training classes. Training sessions should be held regularly, perhaps for a few hours each month, and large-scale exercises should be held once or twice each year. Classroom instruction may include the following:

- Discussion of new ideas, equipment problems, and results of field exercises;
- Movies on new equipment and its use, spill cleanup operations, and drills;
- Status reports on equipment and inventory of supplies;
- Reviews of the emergency contingency plan and responsibilities of individual members; and
- First aid procedures.

Training is time-consuming and expensive. Like any other expenditure, it must be justified by the overall program. Whenever possible, consideration should be given to holding joint sessions with other plants in the area, especially in exercises involving implementation of a large-scale emergency contingency plan.

BIBLIOGRAPHY

American Conference on Governmental and Industrial Hygienists, 1974, *Industrial Ventilation-A Manual of Recommended Practice*, 13th ed., Cincinnati, OH.

_________, 1982, *TLVs: Threshold Limit Values for Chemical Substances and Physical Agents in the Workroom Environment*, Cincinnati, OH.

American National Standards Institute, 1980, *Practices for Respiratory Protection*, New York, NY.

Brief, R.S., 1975, *Basic Industrial Hygiene*, American Industrial Hygiene Associatin, Akron, OH.

Clayton, G.D., and Clayton, F.E., 1978, *Patty's Industrial Hygiene and Toxicology*, Vol. I, 3rd ed., John Wiley and Sons, New York, NY.

Cralley, L.J., and Cralley, L.V., 1979, *Patty's Industrial Hygiene and Toxicology*, Vol. III, John Wiley and Sons, New York, NY.

Maryland Department of Licensing and Regulations, Division of Labor and Industry, 1978, *Maryland Occupational Safety and Health Standard for Confined Spaces*, Baltimore, MD.

United States Department of Labor, Occupational Safety and Health Administration, 1981, *OSHA Safety and Health Standards, General Industry* (29 CFR 1910), OSHA 2206, U.S. Government Printing Office, Washington, DC.

_________, 1979, *OSHA Safety and Health Standards, Construction* (29 CFR 1926/1910), OSHA 2207, United States Government Printing Office, Washington, DC.

United States Department of Health and Human Services, 1980, *NIOSH/OSHA Pocket Guide to Chemical Hazards* 78-120, United States Government Printing Office, Washington, DC.

_________, 1976, *A Guide to Industrial Respiratory Protection*, NIOSH publication 76-189, United States Government Printing Office, Washington, DC.

_________, 1973, *The Industrial Environment-Its Evaluation and Control*, United States Government Printing Office, Washington, DC.

7. Spill Control and Prevention

A pollution event usually consists of an unforeseen occurrence in which a hazardous material is released to the environment. Control of such an event consists of timely and effective response to the spill. Although spills on land, in the air, and in water require different types of response, certain responses are required for any type of spill. To maximize the response team's effectiveness, contingency plans for facility emergencies should be developed and available prior to occurrence of an emergency. Many types of hazardous material spills must be reported promptly to regulatory agencies. There are different reporting requirements which are dependent upon the type of material, volume of material spilled, and location of the spill. These various reporting requirements should be determined by each facility as part of its emergency contingency planning prior to any emergency.

Upon initial notification of a pollution event, response personnel should determine if the source of the spill has been eliminated. Operating personnel at the source are usually most knowledgeable about the cause of the spill and are on hand to take the first steps toward source elimination. Early elimination of the source, whether by diverting a pipeline flow, closing a valve, draining a tank, or any other method, limits the amount of material spilled, and thus the ultimate environmental damage, the cost and effort required for cleanup, and the cost in terms of loss of the spilled material.

At the same time, response personnel must consider the threat to the local human population, and countermeasures must be considered for their welfare. Especially in a hazardous materials storage complex, the danger of a fire or explosion is always a concern. The hazard to the human population must be eliminated even before considering environmental responses. Use of fluorocarbon-water foams to suppress volatilization may be advisable and is just one example of a technique to reduce risk of fire, explosion, or toxic gas spread.

7.1 LAND POLLUTION CONTROL

For spills on land, prompt confinement and removal must be performed to prevent migration of spilled material to surface or groundwater. Groundwater may act as a conduit to other areas, or may be a source of drinking water.

With some knowledge of the local soil properties, response personnel can determine the extent of contamination from a spill, whether groundwater has been contaminated, and the spread of material in the groundwater system. First, the maximum depth of penetration for chemical products, must be determined. The following equation may be applied:

$$D = \frac{K\ V}{A}$$

Where D = maximum penetration depth, in m

V = volume of material entering the ground, in m^3

A = area over which material is spilled, in m^2

K = a constant based on soil retention capacity and viscosity.

If the calculated value for D is greater than the known depth to groundwater, the material has probably penetrated to and spread on the surface of the groundwater. The extent of the spread can be determined by another equation:

$$S = \frac{1000}{F} \left(V - \frac{A\text{-}d}{K} \right)$$

Where S = maximum spread of material, in m^2

F = thickness of chemical layer on groundwater surface, in mm

V = volume of material entering the ground, in m^3

A = area over which material is spilled, in m^2

d = depth to groundwater, in m

K = a constant based on soil retention capacity and viscosity.

Typical values for K and F for some oil products are given in Table 7-1.

7.1.1 Containment Techniques

Upon elimination of the source of a spill, or if manpower permits, concurrently, response personnel should begin containment operations. Although spilled liquids do not migrate on land rapidly, they do seek the lowest areas. For this reason storm drains or other conduits should be blocked as soon as possible to prevent surface water discharges.

Containment measures on land surfaces can vary depending on the amount and type of material spilled, the land gradient, and other factors. If such materials are readily at hand, sorbent booms containing straw or synthetic sorbent can be placed around a spill area. Sand bags can also be used to temporarily hold back spilled material. More sophisticated means of containment include anti-wetting agents to

Table 7-1

TYPICAL K AND F VALUES FOR DETERMINING
EXTENT OF GROUND CONTAMINATION

	K			F (mm)
Soil Type	Gasoline	Kerosene	Light Fuel Oil	
Coarse gravel	400	200	100	5
Gravel to coarse sand	250	125	62	8
Coarse to medium sand	130	66	33	12
Medium to fine sand	80	40	20	20
Fine sand to silt	50	25	12	40

Note: Values given are for relatively light oil products; heavier materials, such as #6 fuel oil or heavy crude oil, are less likely to migrate in the ground due to their greater viscosity.

Source: Texas A & M, 1978.

minimize ground percolation and arrest surface flow; gelling agents, which immobilize the spill by solidification; and imbiber beads, which selectively absorb a wide variety of organic liquids, while allowing water to pass through.

If these readily deployable materials are not on hand, immediate containment must be effected by constructing earthen dikes or berms. For small spills in unpaved areas, personnel can construct berms using hand tools. For larger areas, earthmoving equipment would be more efficient. A backhoe or trench digger can throw up a low berm (30 to 40 cm in height) around a spill area, at the same time creating a shallow trench for collection of spilled materials. On asphalt surfaces, a trench digger or tractor with ripper attachment can be used for berm construction. On concrete paved surfaces, it is more expedient to bring in sorbent booms or sandbags than to rip up the concrete surface.

If spilled material reaches groundwater, containment efforts must be directed toward excavation of an interceptor trench (see Figure 7-1) downgradient of the spill site. Such a ditch can be excavated by a backhoe, if groundwater depth is shallow, or by a clamshell bucket and crane, if the depth is greater than two or three meters. If the water table is too deep for installation of interceptors, pumping wells to locally depress the water table and limit contaminant migration can be installed (see Figure 7-2).

7.1.2 Removal and Treatment Techniques

Removal of spilled materials should be initiated as soon as containment is effected, in order to minimize the amount of contaminated soil and other materials which would also have to be removed. In some instances, the specific removal and treatment technique will require prior approval or concurrence of a regulatory agency. In all cases, health and safety considerations to cleanup personnel and the public must always be addressed when selecting a removal and treatment technique. Health and safety considerations are often regulated, and the appropriate regulation must be consulted for specific requirements. Health and safety is further discussed in Section 6 of this manual. The following paragraphs provide examples of removal and treatment techniques for certain situations. These techniques and individual situations are subject to applicable regulations and agency policies. Field response organizations must secure authorization from the appropriate regulatory body prior to initiation of response activities.

For spills on non-porous surfaces, such as concrete or asphalt, accumulations of material can be vacuumed into holding tanks. Smaller quantities can be blotted with sorbent pads. If elimination of all residues is desired, any material not blotted can be blasted with a high-pressure water hose, and the residue picked up from the water with sorbent pads.

For spills on unpaved surfaces, such as soil or sandy areas, removal of the spilled material from the surface is not possible. In

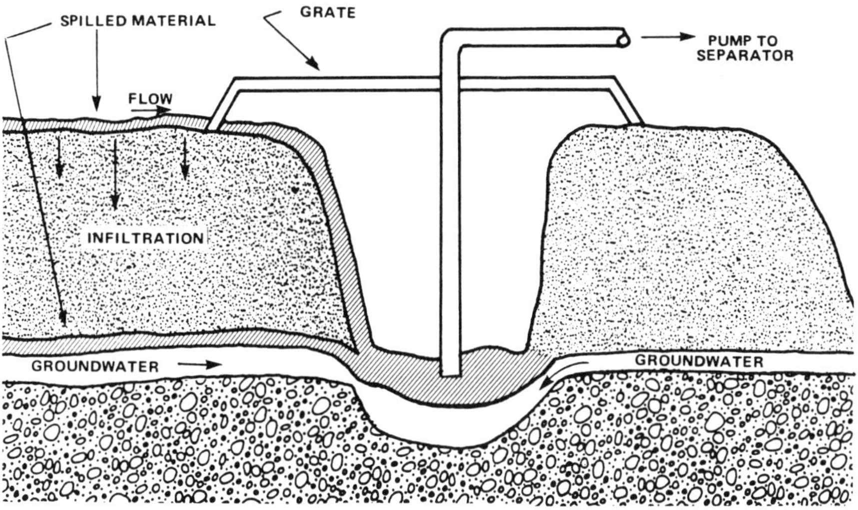

SOURCE: Texas A&M, 1978

Figure 7–1 CROSS-SECTION OF INTERCEPTOR TRENCH CONTAINMENT AND COLLECTION SYSTEM FOR FLOATING CONTAMINANTS

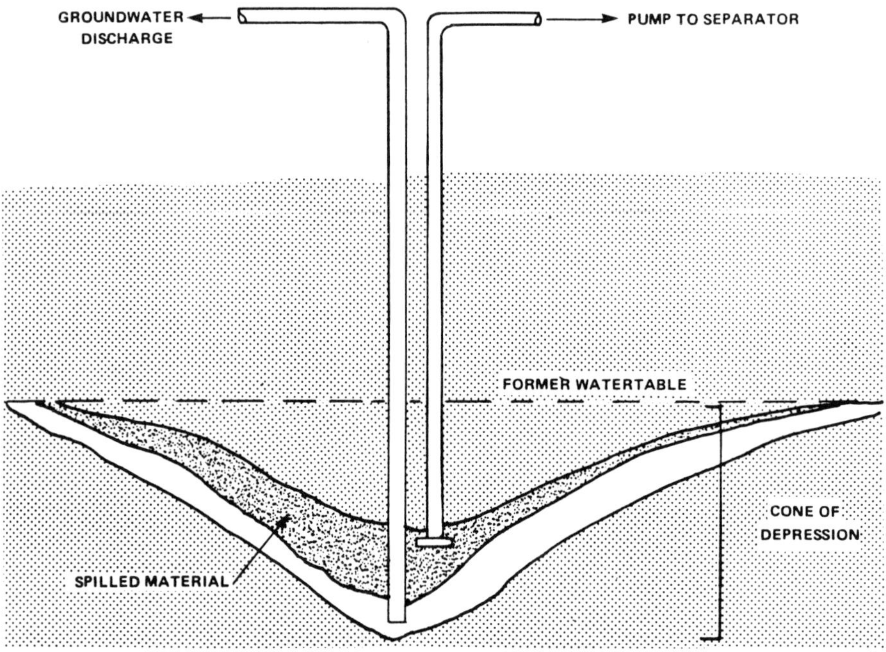

SOURCE: Texas A&M, 1978

Figure 7–2 SCHEMATIC OF DEEP GROUNDWATER RECOVERY WELL FOR FLOATING CONTAMINANTS

such spills, excavation of the contaminated medium is necessary. Excavation can best be performed by front-end loaders or scrapers, with the contaminated soil then being carried away by dump trucks. For small spills, hand-shoveling can be used to remove the contaminated soils.

In some areas, access for heavy equipment may be extremely limited, and yet the material may be too extensive for manual removal. In such situations, microbiological agents may be applied to the spill area. These agents generally consist of a bacteria adapted to feed on the material and a starter culture medium. Water may have to be treated with the bacterial agents.

In some cases, hazardous materials are best treated on the spill site. Inorganic acids can be neutralized by the addition of lime, and inorganic bases by the addition of dilute acetic acid, and then discharged into a wastewater treatment facility. Hypochlorites and other strong oxidizers can be reduced with sodium sulfite and discharged to a wastewater treatment plant. Sodium sulfide can be used to precipitate heavy metals from spilled fluids.

Treatment of organic hazardous materials at the site of a spill is a more complex problem. Specialized process equipment must be brought to the spill site. Portable carbon filters, wet air oxidation apparatus, or chemical fixation units could be prepared in advance. Appendix D contains chemical treatment information regarding many specific chemicals.

Removal of spilled material from groundwater often involves the use of separators and holding tanks. Skimming devices and pumping systems specially designed and constructed for the materials being handled should be utilized whenever possible. When this equipment is not available, the contaminated groundwater can be pumped into collection trenches or pools equipped with an impermeable liner or barrier. Vacuum equipment, such as a surface skimmer or pump and hose, may then be used to separate material in the trench or pool. In some cases, material from the trench or pool could be pumped into a gravity separator erected on-site. Water from the separator can be discharged to a wastewater treatment plant. The oil or chemical fraction could then be drained for proper recycling or disposal.

There are a variety of other removal systems for spills reaching groundwater. Location and shape of the trench and separation mechanisms on-site should conform to the requirements of the particular spill event. Spills of materials denser than water or miscible with water require a deeper trench and a different separation or treatment process. Some situations may warrant construction of wells rather than trenches, and considerable pumping from the wells. The possibilities are numerous and require case-by-case evaluation by response and regulatory personnel.

7.1.3 Disposal Techniques

Materials contaminated with oil or hazardous chemicals should be disposed of only in treatment or disposal facilities that are designed and permitted by applicable regulatory agencies to handle such materials. Particular care should be taken that all wastes generated by cleanup of a hazardous material spill be disposed of in accordance with all applicable state or federal hazardous waste regulations.

In most instances, the particular method of treatment and disposal of spill debris must be determined on an individual basis. References on the techniques available can be found in the bibliography. Facility contingency plans (Section 7.5) and careful preplanning should include identification of hazardous waste transporters, disposers, and cleanup companies that would be available in the event of an emergency.

7.2 AIR POLLUTION CONTROL

Oil or hazardous material pollution situations affecting the air mostly occur in terms of releases of hazardous vapors. Such emissions may range anywhere from a chronic, low-concentration leak from a vapor trap, to a widespread expanding vapor cloud. These situations must be handled on a case-by-case basis. However, general methodologies can be presented for consideration.

7.2.1 Local Meteorology

Pathways of distribution of vapor releases are determined by local meteorological conditions. These conditions vary, even on a day-to-day basis, such that meteorological information must be determined for the vapor release before proper response activities can be identified. Such information can be obtained from existing local air-monitoring stations.

The most significant local factor to consider in hazardous vapor releases is wind conditions. Atmospheric stability, which is influenced by wind speed, is also of considerable concern in the distribution of vapor releases.

7.2.2 Air Emissions Control

Upon notification of a discharge, the initial response should be directed toward eliminating the source of the discharge. Due to the hazardous nature of vapor discharges to the atmosphere, response personnel may have to wear protective equipment. Because of toxic, corrosive, or reactive vapors, personnel may also have to use an alternative air supply and protective, or even isolating, garments for operations near the source of the emission. Operating personnel can usually verify the nature of the material. The quantity of materials released can be estimated and the toxicity of the materials can be learned.

Personnel charged with eliminating vapor discharge sources should have available to them a variety of plugs and patches to stop tank, transfer line, and pipeline leaks. Such devices should include expandable rubber and polyethylene plugs of various sizes for round holes or those that could be rounded, and patch materials, such as aluminum plates, heavy rubber mats, bands, and other holding devices for tears and other odd-shaped ruptures.

If the emission source cannot be eliminated, it must be controlled. For containers of liquefied flammable gas in or near a fire, the possibility of a Boiling Liquid Expanding Vapor Explosion (BLEVE) exists. Containers of such materials must be kept cool, usually by spraying them with a continuous stream of water or a heat-absorbing foam. Less volatile materials escaping as fumes or mists may be knocked down with a water spray or firefighting foams. Water containing fume or mist products should be collected and treated as a land pollution situation, as outlined in Section 7.1.

Other techniques for control of air emissions include the use of vents to divert a vapor to a holding container and then to a treatment facility. A pipe vent can be installed on a tank or pipeline. For vapors denser than air, a trench vent may have to be dug with a backhoe or trench-digger, and the vapor can then be treated or stored below ground level.

In some situations it may be necessary to place an emergency structure around and over the emission source. For small, isolated problems, such as leaking drums, this technique is easily applied using overpack drums. Some situations may require incorporating mitigation measures into the design.

7.2.3 Vapor Emission Treatment

Once vapor emissions are under control, available measures to treat the vapors should be used. For tanks located at an industrial facility, it may be possible to collect and route controlled vapor leaks to the plant's air pollution control equipment. Collected organic vapors can be routed to the appropriate liquefaction or vapor recovery system. Acid gases can be fed to a scrubber or precipitator with the operation's exhaust gases. Other treatment techniques such as water sprays, foam blankets, and cryogenic techniques can be used to control vapor emissions. In some cases, normal atmospheric dispersion must be relied on for returning the air to a safe, breathable level. In confined areas or in very localized situations, blower fans may assist in dispersing air contaminants.

For more chronic air pollution problems, such as waste disposal sites, extraordinary gas treatment, such as vapor phase adsorption or thermal oxidation, may be required.

7.3 SURFACE WATER POLLUTION CONTROL

The materials available for control of water pollution events, and their applications, vary considerably. Most control equipment and applications are for oils and chemicals that behave like oils, i.e., that are relatively insoluble in water and are less dense than water, and so float. Appendix D refers to appropriate land and water spill control and cleanup techniques that should be used for a wide variety of specific chemicals.

Containment of spills must be prompt. In open water, even a small spill may spread out to affect a large area, which would require extensive cleanup. Table 7-2, from the United States National Oil and Hazardous Substances Pollution Contingency Plan, describes the visual appearance of various quantities of oil on water.

In addition to the type and approximate quantity of material spilled, response personnel must consider the spill location, wind speed and direction, and tide and current speed and direction, in order to select the proper response materials and techniques.

7.3.1 Control Equipment

Control equipment consists of materials used for containment of a spill, such as booms, air compressors, or hoses and pumps, coupled with removal equipment, such as skimmers, sorbents, dredges, or earth-moving equipment. In certain situations, chemicals can be used in lieu of control equipment.

7.3.2 Containment Equipment

Booms are the primary devices used for containment, either in open water or along shorelines. Although there is a wide variety of boom designs, the basic containment boom (see Figure 7-3) contains the following properties or components: enough buoyancy to keep the boom above the surface and prevent the spilled material from slopping over the boom; a skirt to collect the material and prevent it from drifting under the boom; weight to maintain proper orientation to the accumulated material; and points of attachment, for extending the boom by attaching additional sections and for tethering the boom to an anchor float (see Figure 7-4).

Application of booms to spill situations requires personnel trained in their use and limitations. Booms are limited to use in currents less than 1.3 knots. At velocities greater than 1.3 knots, oily material is entrained under the skirt. In order to reduce current velocity with respect to the boom to 1.3 knots, the boom must be deployed at an angle to the direction of the current. Given the current velocity, using Table 7-3, response personnel can deploy a boom at the proper angle. A boom deployment of about 70° is the maximum angle recommended.

Boom configurations vary with the situation. In open water a boom can be used to encircle an oily material and confine it for removal. Along shorelines, booms can be angled, depending on water and wind velocities, to channel a spill to collection points or away from sensitive areas.

Booms used in sea areas and where strong wind-driven currents occur should be of the heavy-duty type. Such booms usually have a freeboard of 18 inches and a draft of 24 inches. For shallow water nearshore areas, booms with 6- to 10-inch freeboards and up to 12-inch drafts are more useful, since they can be brought closer to shore and still retain their effectiveness.

Table 7-2

VISUAL APPEARANCE OF VARIOUS QUANTITIES OF OIL ON WATER

Standard Term	Gallons of Oil Per Square Mile	Appearance
Barely visible	25	Barely visible under most favorable light conditions.
Silvery	50	Visible as a silvery sheen on surface water.
Slightly colored	100	First trace of color may be observed.
Brightly colored	200	Bright bands of color are visible.
Dull	700	Colors begin to turn dull brown.
Dark	1,300	Colors turn a much darker brown.

Note: 1. The terms used to describe an oil film, which is a slick thinner than 0.0001 inch, are given below (Council on Environmental Quality 1979).

2. Each 1-inch thickness of oil equals 5.6 gallons per square yard, or 17,000,000 gallons per square mile.

Source: USEPA 40 CFR 112.

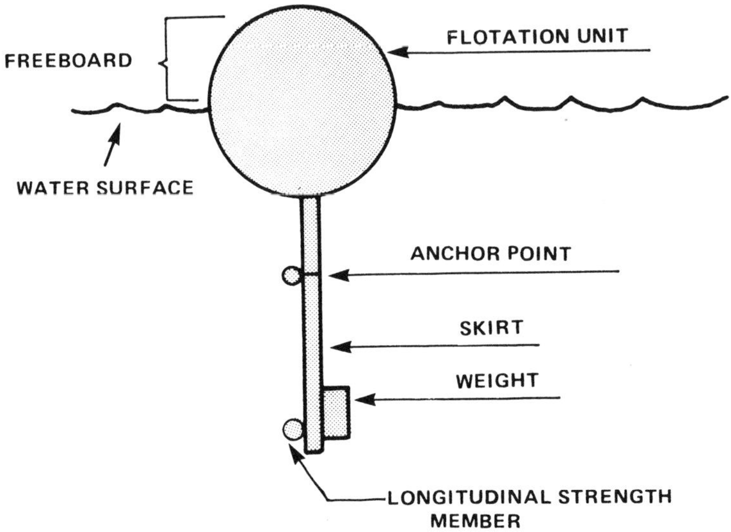

SOURCE: Texas A&M, 1978

Figure 7–3 CROSS-SECTION OF A TYPICAL BOOM, SHOWING MAJOR PARTS

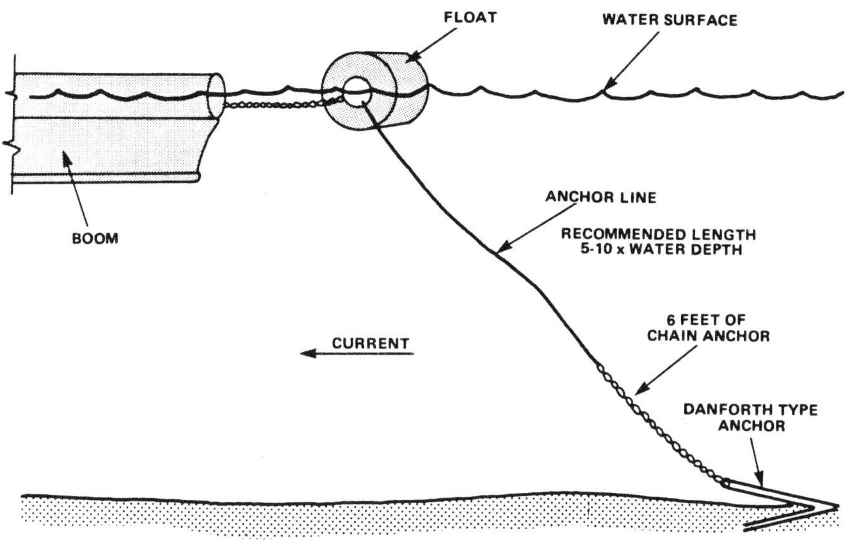

SOURCE: Texas A&M, 1978

Figure 7—4 SCHEMATIC OF TYPICAL BOOM ANCHORING SYSTEM

Table 7-3

BOOM ANGLES FOR FLOW VELOCITIES GREATER THAN 1.3 KNOTS

Velocity in Knots	Velocity in m/sec	Angle to Reduce Velocity to 1.3 Knots
1.5	0.76	30°
1.6	0.81	35°
1.7	0.86	40°
1.8	0.93	45°
2.0	1.02	50°
2.3	1.15	55°
2.6	1.32	60°
3.1	1.56	65°
3.8	1.93	70°

Source: Ecology and Environment, Inc., 1982.

Sorbent booms are another type. They consist of sorbent material, usually synthetic, stuffed inside elongated plastic mesh bags. Having no skirts, they have negligible draft and thus function best as polishing devices downstream or downcurrent of the primary containment booms. They are suitable for containment when necessary.

Underflow dams are commonly used in small streams to contain materials that float on the water surface (see Figure 7-5). Underflow dams are commonly used in small streams. These dams consist of dikes with angled pipes or outlet structures through them that release water from the bottom of the stream. This allows clean water to be discharged, while surface contaminants are retained behind the dam.

Another containment technique, restricted to calm water, consists of perforated pipe or hose through which air is forced from an onshore compressor. The pipe is laid on the channel bottom, and the curtain of rising bubbles from the pipe creates a double vortex movement in the water, separating the waters on either side.

For substances that are heavier than water or miscible with water, there are few viable containment techniques. In many instances, removal or treatment methods are used, rather than containment.

For materials that are heavier than water, overflow dikes constructed of earth or other materials can be used for containment. For materials that are miscible with water, filter fences, stream diversion systems, or damming the total stream flow can be utilized.

7.3.3 Removal Equipment

Boom containment is best used as a temporary holding operation. Without prompt removal of material, entrainment and shifting winds and tides will allow collected material to drift away from the boom. Thus, equipment for the removal of spilled product should be deployed as soon as possible.

Skimmers or vacuum equipment are most commonly used for removal of significant concentrations of spilled materials. Skimmers work on one of several principles: gravity, suction, or adhesion.

Gravity skimmers basically consist of a float unit with an overflow weir which is adjusted to ride at the water surface, so that floating material passes over the weir into a reservoir (see Figure 7-6). From the reservoir the material is pumped away for treatment or disposal. Advancing weirs and double advancing weirs, which allow release of collected water from the reservoir, are variations of the basic floating weir. Weirs are very mobile and easily deployed, but with any wave action easily lose their efficiency. They are best used in calm, debris-free waters such as channels and boomed ship berths.

Suction skimmers vacuum the spilled material from the water either by means of one or more broad, floating vacuum heads (see Figure 7-7), or by creating a water vortex, which pulls surface

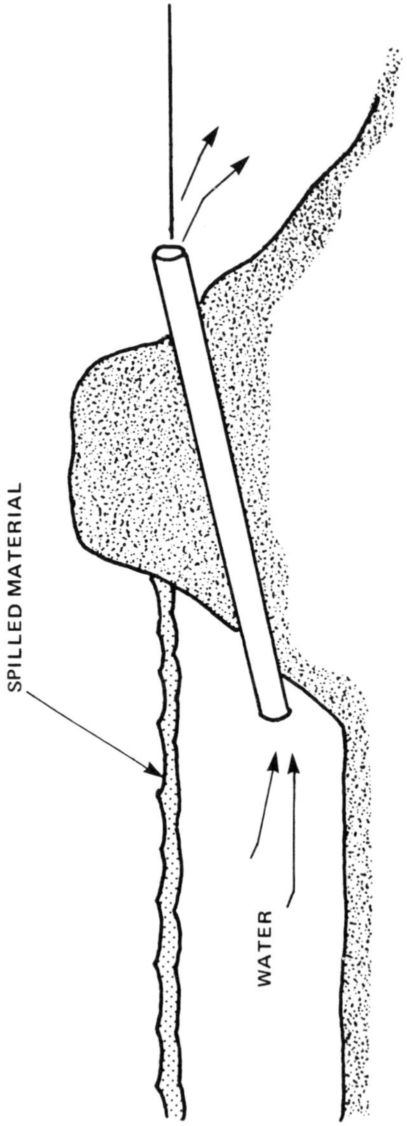

SOURCE: Texas A&M, 1978

Figure 7–5 SCHEMATIC OF TYPICAL UNDERFLOW DAM

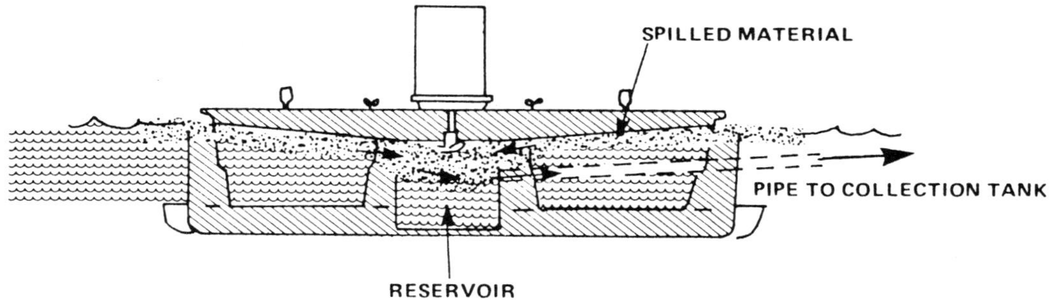

SOURCE: Texas A&M, 1978

Figure 7–6 CROSS-SECTION OF TYPICAL FLOATING WEIR SKIMMING UNIT

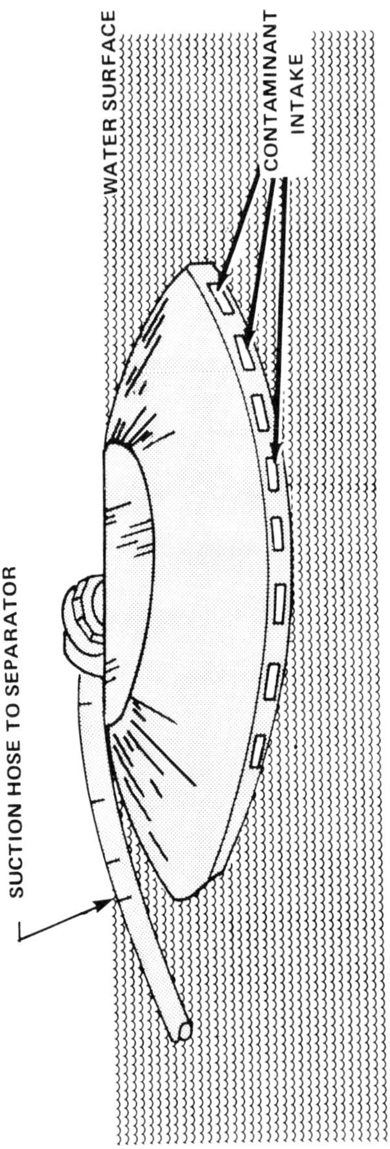

SOURCE: Texas A&M, 1978

Figure 7–7 ILLUSTRATION OF FLOATING SUCTION SKIMMING UNIT

material down into a whirlpool. Vacuum hoses then carry the material to holding tanks for recycling or disposal. Suction skimmers can have many types of vacuum pumps, including rotary gear, rotary vane, centrifugal, cloverleaf, and internal gear pumps. Such pumps can be coupled to a thick-walled vacuum hose for applications to a wide variety of situations. Suction devices can be compact, yet have a high capacity. Floating head units can be used in very shallow water, but currents must be fairly slow (less than 0.6 knots) to prevent the units from planing over the water surface. Suction units work well in inshore areas where booms are used to confine the spilled materials.

Adhesion equipment includes both rotating drum (see Figure 7-8) or nonporous collecting belt devices and oleophilic ropes and belts. The nonporous belts pull product under the water and cause it to surface in a collection well (see Figure 7-9). The oleophilic belts and ropes adsorb product, after which it is squeezed off rollers into a collection container (see Figures 7-10 and 7-11). The adhesion units are highly efficient, especially in rough water or where floating debris is present. While the oleophilic materials work best with medium-viscosity oily materials, the nonporous belts are capable of handling a wide range of viscosities, including lightweight materials. Since these units are somewhat large and require trained personnel close at hand, they are best suited for operation on the bow of a recovery boat. Such arrangements are useful for open sea or open harbor situations.

While skimmers or vacuum equipment are best suited for initial removal of accumulations from a water surface, sorbent materials are often necessary for a final cleanup. Sorbents are also useful in situations where booms and skimmers are difficult to operate, such as under wharves or in areas with much debris. Some properties of sorbents are described in Table 7-4. Use of loose sorbents should be avoided for large spills, as retrieval of them is manpower-intensive. Sorbent booms are useful in the protection or isolation of sensitive areas or areas difficult to clean up, as well as for normal containment. For substances which are heavier than water or miscible with water, dredges and filter fences can be utilized. There is a wide variety of dredges which can be used to remove materials from the bottom of waterways for subsequent disposal and treatment. A wide variety of filter media can be used to filter water-miscible contaminants from a flowing waterway or water column. The use of dredges and filter fences should be determined on a case-by-case basis, with considerable care and judgement exercised in order not to create a greater problem than already exists. In many cases of spills of materials that are heavier than or miscible with water, there is little that can be done to remove the contaminants.

Other removal materials may be required for cleanup of contaminated shorelines. Steam-cleaning equipment can be used on bulkheads and pilings if sorbent booms are used to trap spilled materials. High- or low-pressure water spray equipment can be used to displace materials from some shorelines, such as sand beaches. Otherwise, heavy equipment, such as bulldozers, scrapers, or backhoes, may be required, with the contaminated materials removed for subsequent treatment and disposal.

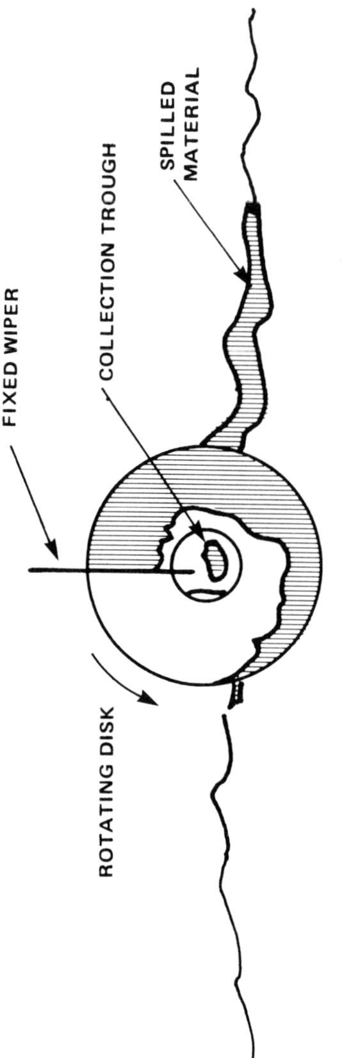

SOURCE: Texas A&M, 1978

Figure 7–8 CROSS-SECTION OF TYPICAL OLEOPHILIC DRUM SKIMMER

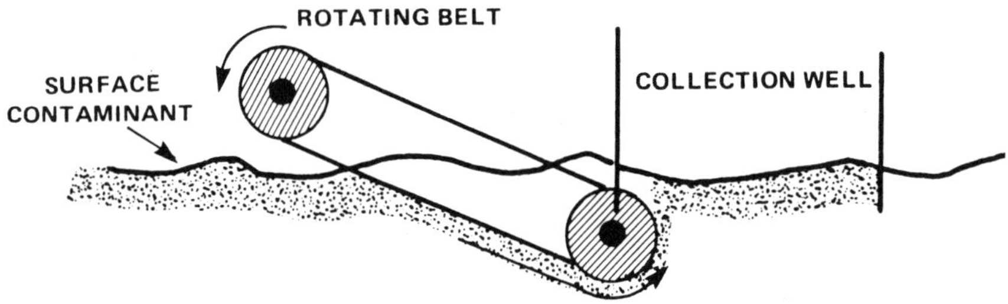

SOURCE: Texas A&M, 1978

Figure 7–9 SCHEMATIC OF INCLINED PLANE BELT SKIMMER

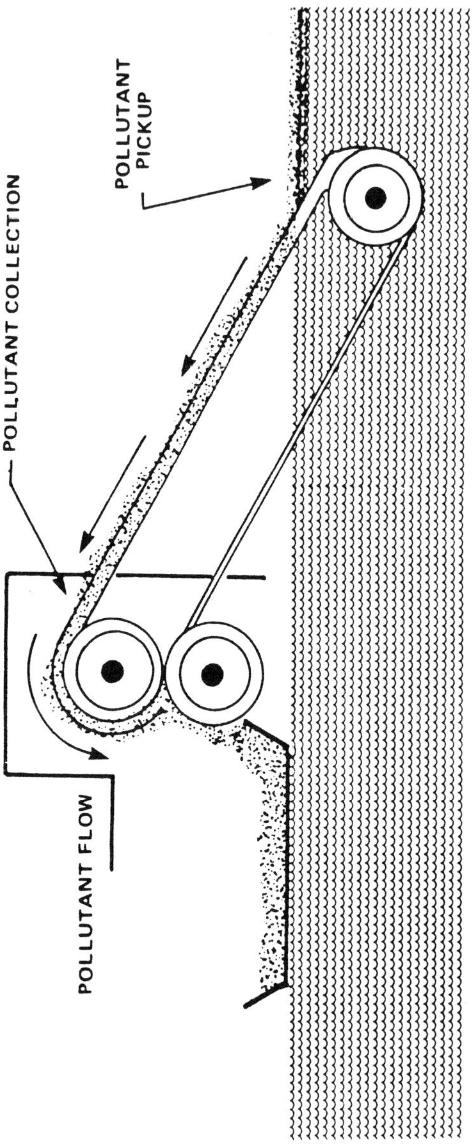

SOURCE: Texas A&M, 1978

Figure 7–10 SCHEMATIC OF OLEOPHILIC BELT SKIMMER

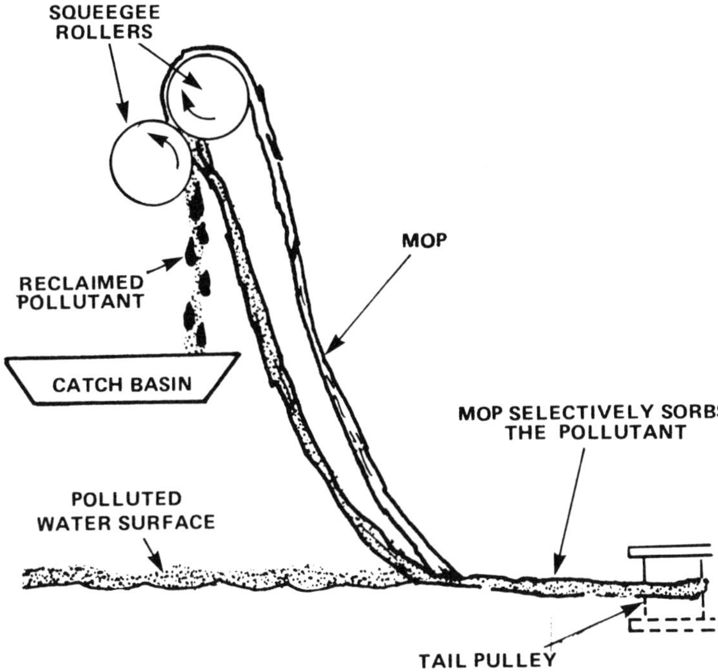

SOURCE: Texas A&M, 1978

Figure 7–11 SCHEMATIC OF OLEOPHILIC ROPE SKIMMER

Table 7-4

PROPERTIES OF SORBENT MATERIAL

Type	Advantages	Disadvantages	Example	Capacity
Natural Sorbents	Non-toxic, biodegradeable	Soak up both organics and water; will sink when saturated	Peat moss	In general, absorb 3 to 6 times their weight
			Straw	
		Recovery of large amounts of sorbent is a labor-intensive operation	Milled corn cobs	
		Trapped product may drain off sorbent material	Wood cellulose fiber	
			Milled cottonseed fiber	
Inorganic or Mineral-Based Sorbents	Relatively inexpensive	Very light materials; difficult to distribute when windy	Perlite	In general, absorb 4 to 8 times their weight
		Non-biodegradable	Vermiculite	
		Dust may cause respiratory irritations	Volcanic ash	
		Can be abrasive to recovery equipment		
Synthetic Sorbents	Exceptionally high recovery efficiences	Expensive	Polyurethane	Variable, but higher than non-synthetic solvents, typically about 20 to 25 times their own weight
	Some materials can be re-used after oil removal	Non-biodegradeable	Urea formaldehyde	
			Polyethylene	
	Easily spread		Polypropylene	

Table 7-4 (Cont.)

Type	Advantages	Disadvantages	Example	Capacity
	Easily recovered Available in many forms (e.g., rolls, sheets, booms)			
Synthetic Foam Sorbents	Most efficient sorbents available Efficiency independent of viscosity Can be produced on-site by mixing two liquids	Saturated slabs may tear during recovery	Polyurethane foam	Variable, but higher than non-synthetic solvents, typically about 20 to 25 times their own weight

Source: <u>Handbook for Oil Spill Protection Cleanup Priorities</u>, 1981, Versor, Inc.

Although physical removal measures are usually preferable, chemical and biological agents for spill control are available and are also used. These include burning agents, sinking agents, biological cultures, collecting agents, and dispersants. Appendix D identifies appropriate treatment and removal methods for an extensive list of chemicals.

In all removal instances, the specific situation at hand must be evaluated and action taken that is often innovative and unique to the given situation. Experienced and competent response personnel must be imaginative and resourceful to effectively handle a spill within limited time and resources constraints.

7.4 SPILL PREVENTION, CONTROL, AND COUNTERMEASURE PLANS

Spills of oil and hazardous substances and their effects can be minimized with proper implementation of a Spill Prevention, Control, and Countermeasure (SPCC) Plan. An SPCC plan provides a comprehensive system of spill prevention which includes requirements for secondary containment, inspection and maintenance, facility drainage, facility transfer operations, facility security, and other aspects of a facility which are related to spills, and personnel training. An Emergency Contingency Plan which delineates procedures to be taken in the event of an uncontained spill or other emergency should be used in conjunction with an SPCC plan. Such a program not only protects human and environmental health but also protects a facility from the huge financial repercussions which almost always accompany emergency clean-up operations.

The initial stage of SPCC program development involves writing a plan with the aid of a Professional Engineer. Once the plan is written, it should be present at the facility at all times. A proper SPCC plan must include:

- The name, type, location, and start-up date of the facility to which it applies;
- Identification of the facility owner/operator and the person responsible for spill prevention at the facility;
- Management approval, usually indicated by signature;
- The name and certification of the Professional Engineer who prepared the plan;
- Spill history and prediction of potential spills; and
- Descriptions of secondary containment, drainage, storage tanks, transfer operations, inspection and maintenance procedures, security, and personnel training at the facility.

SPCC plans should always be kept up-to-date. Management reviews must be conducted at least every three years. The plan must be reviewed and, if necessary, amended and recertified whenever the facility is modified. Additional guidance on SPCC planning can be found in API Bulletin D16, and EPA regulation 40 CFR Part 112. These guidelines refer specifically to oils but are applicable to hazardous materials as well.

7.4.1 Spill History and Prediction

A facility which has experienced a spill must include in the plan a description of the spill which identifies the chemical and amount spilled; the location, date, and time of the spill; any water source affected; the related damages and costs to the facility; the cause of the spill; and action taken to prevent recurrence.

All SPCC plans must include an analysis of the facility which indicates the magnitude of spill potential. Sources should be identified as to the chemical and amount stored, and a prediction of the direction of spill flow should be included.

7.4.2 Secondary Containment

A description of the facility's secondary containment scheme must be included in every plan. Guidelines for adequate secondary containment have been discussed in Sections 3.8, 7.1, 7.2, and 7.3 of this manual.

7.4.3 Facility Drainage

Every plan must include a description of drainage systems employed at the facility. Secondary containment areas which require drainage should contain valves with manual open and close design. These areas should be drained off-site only after the drainage water has been analyzed. If drainage is to an in-plant treatment facility, that facility should be designed to handle any chemical which the drainage might contain. Plant drainage from undiked areas should flow to a collection point or there should be a diversion system that can return the drainage to the facility in the event of a spill.

7.4.4 Tanks

The plan must confirm that tank contents are compatible with the tank construction material and that tanks and tank supports are of sound design. There should be some means of preventing overfilling of tanks, such as a liquid level indicator, high level alarm, or pump cut-off device. Underground tanks and pipelines should have adequate corrosion protection, and the method employed should be described in the SPCC plan.

7.4.5 Facility Transfer Operations

Every plan should insure that pipe supports are of adequate strength and design and that aboveground pipelines are protected from vehicular traffic. Pipelines not in service must be capped or blank-flanged. Potential leak spots, such as at valves or joints, can be protected with drip pans or some other method.

For transfer operations conducted between sunset and sunrise, a minimum lighting intensity of five foot-candles should be provided for the transfer connection points, and all work areas should have a minimum lighting intensity of one foot-candle.

7.4.6 Inspection and Maintenance

All tanks, tank supports, pipelines and secondary containment structures must be inspected regularly for integrity. Tanks should be inspected daily for loss of product and integrity-tested at least once every five years. Integrity testing can be done visually, with hydrostatic testing, or by some other non-destructive method. All inspection methods and schedules must be described in the plan. Inspection and maintenance records dating back at least three years must be kept at the facility. Inspection and maintenance procedures are discussed in Chapter 5 of this manual.

7.4.7 Security

Because of the hazardous nature of the materials that will be stored, all plans must include facility security. The temporary storage area must be designed to prevent the unauthorized entry of persons, vehicles, or animals. This may involve the construction of fences, walls, or an impassable ditch around the area. Warning signs should also be posted around the perimeter of any storage area. These signs should be large enough to be easily read from a distance of at least 25 feet and should include an international symbol to warn of danger, as well as warnings (in English and any other language appropriate to the area) to unauthorized persons to keep out of the area. All entrance gates should be locked when the facility is unattended. All valves and pump controls should be locked when not in use. Adequate lighting of the facility must be provided.

7.4.8 Personnel Training

Appropriate facility personnel should be familiar with the SPCC plan. Spill prevention meetings should be held at least once per year to train or retrain personnel. Guidelines for personnel training are presented in Section 6.

7.4.9 Other Considerations

Many facilities have features that require special consideration in SPCC planning. For instance, the nature of the chemicals stored may be such that, in the event of a spill, an open drainage collection basin such as a pond or lagoon could cause an immediate threat to facility employees from vapor exposure. In such a situation, a closed system would be required. Special considerations are required when incompatible chemicals are stored at the same facility. For example, secondary containment should be constructed such that tanks of incompatible materials are isolated from each other. Location of a facility can strongly influence SPCC planning, as in the case of facilities situated in floodplains. Management commitment to spill prevention and thoughtful SPCC planning in general should insure that special considerations are adequately addressed in an SPCC plan.

7.5 EMERGENCY CONTINGENCY PLANS

The quicker a chemical spill or other emergency is attended to, the smaller the threat to human and environmental health, and the less costly the remedial action. Facilities can insure prompt emergency response by preparing a comprehensive Emergency Contingency Plan that contains information on emergency equipment, procedures, and sources of assistance.

The following general requirements should be included in contingency planning:

- All plans should be regularly reviewed and, if necessary, amended;
- Appropriate personnel should be familiarized with the plan at least annually;
- Amendments are required whenever the facility permit is revised, if the plan fails in an emergency, or if there is a change in key personnel;
- An inventory of emergency equipment is required in the plan;
- Sources of assistance should be identified; and
- Emergency procedures must be described in detail.

Additional guidelines on contingency planning can be found in EPA regulations 40 CFR, Parts 264, 265, and 300. A comprehensive emergency control program can best be achieved by developing an Emergency Contingency Plan in conjunction with an SPCC plan.

7.5.1 Emergency Equipment

An inventory of equipment for fire control, spill control, and decontamination should be included in the plan. This list should indicate the location and general capabilities of each piece of equipment. Some key pieces of equipment are booms, sorbent materials, detoxifying materials, firefighting equipment, alarm systems, and emergency telephones. Alarms, telephones, or other communication devices should be so located that they can be easily reached in an emergency. All emergency equipment should be regularly tested and inspected, and appropriate records should be maintained. In addition, an adequate water supply should always be maintained for use in conjunction with emergency equipment.

7.5.2 Sources of Assistance

Various agencies which should be contacted in the event of an emergency should be identified in the plan. The list of assistance sources should include, at a minimum, local police, fire departments, hospitals, spill contractors, and state and local emergency response teams. Identification of these agencies should provide at least the facility name and phone number. In addition to being listed, these agencies should be given copies of the contingency plan as soon as it is completed and should be kept up-to-date on changes in the plan. Other telephone numbers which should be listed are those for the National Response Center, the nearest U.S. Coast Guard Station, and the EPA's local On-Scene Coordinator. An emergency phone numbers form is provided in Figure 7-12.

Key personnel within the spill response organization should be explicitly identified as to their emergency roles. This may be done internally or through a convenient spill contractor. The idea is to organize a spill control team as expeditiously as possible. The plan should include a section on manpower that could be assigned in an emergency.

7.5.3 Emergency Procedures

Every facility should designate one person and several alternates to assume emergency coordination responsibilities. These people should be listed in a ranked order in the plan, and one of these people should always be at the facility or on call. Procedures which the emergency coordinator must oversee should follow a logical sequence. They should include, but not be limited to, the following:

1. Activate alarms or other communication system to alert facility personnel;

2. Organize the in-house response team or notify the local spill contractor;

3. Notify appropriate state and local agencies (which should already have copies of the contingency plan);

4. Characterize the emergency with respect to the source, the amount of released material, and the hazards created;

5. If evacuation is warranted, initiate evacuation procedures (if evacuation involves surrounding areas, outside authorities may be needed to assist);

6. If areas outside the facility are affected, notify the National Response Center or the local On-Scene Coordinator;

7. Take all reasonable measures to keep the spill or fire from spreading;

8. If the facility must halt operations, monitor tanks and pipes for leaks, pressure build-up, gas generation, and ruptures;

9. Provide for treating, storing, or disposing of contaminated soil, water, or other material;

10. Notify state and local officials when the facility is cleaned up and ready to resume operations; and

11. Record the time, date, and details of the emergency.

EMERGENCY PHONE NUMBERS

1. In-house Emergency Response Coordinator and Alternates

   A. Name: ____________ Telephone: ________
      Address: ____________
   B. Name: ____________ Telephone: ________
      Address: ____________
   C. Name: ____________ Telephone: ________
      Address: ____________

2. U.S. Coast Guard:
   Local Phone: ____________

3. National Response Center: (800) 424-8802

4. EPA On-scene Coordinator: ____________

5. State Emergency Government: ____________

6. Local Emergency Government: ____________

7. Hospital/Health Treatment: A. ____________
   B. ____________
   C. ____________

8. Police: A. ____________
   B. ____________
   C. ____________

9. Sheriff: ____________

10. Fire Department: ____________

11. Spill Clean-up Contractors:
    A. ____________
    B. ____________

12. Other:
    ____
    ____

SOURCE: Ecology and Environment, Inc., 1983.

Figure 7–12 EMERGENCY PHONE NUMBERS FORM

Emergency procedures will vary according to when the spill occurs. Appropriate actions can be derived by examining all possible situations. The following factors should be considered:

- Flow conditions in nearby watercourses;
- Time of spill: how fast can emergency procedures be initiated during normal working hours, at night, on a weekend?
- What sensitive environmental areas might be threatened in each situation?

7.5.4 Emergency Data Sheets

Emergency Data Sheets should be completed for every chemical stored. These sheets should include tank and chemical identification information and a brief summary of health effects, fire protection methods, hazardous properties, storage requirements, environmental protection requirements, personnel protection requirements, and any other information about the chemical which would be important in the event of a spill. An Emergency Data Sheet form is provided in Figure 7-13.

EMERGENCY DATA SHEET

Facility ______________________ Telephone __________

Address ______________________________________

____

1. Tank Identification

   A. Tank Number ______________________

   B. Location ______________________

2. Chemical Identification

   A. Name ______________________

      Synonyms ______________________

   B. Molecular Formula ______________________

   C. Molecular Weight ______________________

   D. Boiling Point ______________________

   E. Density ______________________

   F. US DOT Classification ______________________

   G. US DOT I.D. Number ______________________

      CAS I.D. Number ______________________

References:

   NIOSH Registry of Toxic Effects p. ___

   CRC Handbook of Chemistry p. ___

   49 CFR 100-199

   Hazardous Materials Emergency Response Guidebook, US DOT

3. Health Effects

   A. Acute ______________________

   ____

   B. Chronic ______________________

   ____

SOURCE: Ecology and Environment, Inc., 1983.

Figure 7–13 EMERGENCY DATA SHEET FORM

C. Toxicity ______________________________

D. Route of Exposure

Eye Ingestion
Lung Skin

E. First Aid ______________________________
____
____

F. Medical Monitoring ______________________________
____
____

References:

Pocket Guide to Chemical Hazards, NIOSH/OSHA

ACGIH TLV Handbook, Dangerous Properties of Industrial Materials, Sax.

4. Fire Protection

A. Prevention Technique ______________________________

B. Extinguishing Agents ______________________________

C. Combustion Products ______________________________

References:

Fire Protection Guide on Hazardous Materials, NFPA

Hazardous Materials, US DOT

5. Hazardous Properties

A. Major Chemical Incompatibilities ______________________________

References:

CHRIS, Condensed Guide to Chemical Hazards, USCG

Merck Index

SOURCE: Ecology and Environment, Inc., 1983.

Figure 7–13 EMERGENCY DATA SHEET FORM (Cont.)

6. Methods of Storage

A. Primary ______________________

B. Second Containment ______________________

C. Storage Hazards ______________________

References:

Fire Protection Guide on Hazardous Materials, NFPA

7. Environmental Protection

A. For Material in Fire:

____
____
____
____
____
____

B. For Material not in Fire:

____
____
____
____
____
____

References:

CHRIS, Condensed Guide to Chemical Hazards, USCG

Hazardous Materials, Emergency Response Guidebook

Fire Protection Guide on Hazardous Materials, NFPA

Chemtrec, (800) 424-9300

8. Personal Protection

____
____
____
____

SOURCE: Ecology and Environment, Inc., 1983.

Figure 7–13 EMERGENCY DATA SHEET FORM (Cont.)

References:

Fire Protection Guide on Hazardous Materials, NFPA

CHRIS, Condensed Guide to Chemical Hazards, USCG

Hazardous Materials Emergency Response Guidebook, US DOT

Bests' Safety Directory

9. Other Information

____

____

____

____

SOURCE: Ecology and Environment, Inc., 1983.

Figure 7–13 EMERGENCY DATA SHEET FORM (Cont.)

BIBLIOGRAPHY

American Institute of Plant Engineers, Spill Prevention and Control: An Overview for Industrial Plants, International Plant Engineering Proceedings; Anaheim, California, paper 12-B.

American Petroleum Institute, 1978, Waste Oil Roundup, API Publication 1587, American Petroleum Institute Publishers, Washington, D.C.

__________, 1977, Recommended Practice for Bulk Liquid Stock Control at Retail Outlets, API Publication 1621, American Petroleum Institute Publishers, Washington, D.C.

__________, 1971, Recommended Good Practice for Bulk Liquid - Loss Control in Terminals and Depots, API Bulletin 1623, American Petroleum Institute Publishers, Washington, D.C.

American Insurance Association, 1976, Fire Prevention Code.

Berry, R.E., 1980, First Response Procedures for Hazardous Materials, Pollution Engineering, October 1980: 37-41.

Control of Hazardous Material Spills, 1978, Proceedings of National Conference on Control of Hazardous Materials Spills, Information Transfer Inc., Rockville, MD.

__________, 1980, Proceedings of National Conference on Control of Hazardous Materials Spills, Vanderbilt University, Nashville, TN.

Fawcett, H., and W. Wood, 1965, Safety and Accident Prevention in Chemical Operations, Wiley-Interscience, New York, NY.

Goodier, J.L. et al., 1971, Spill Prevention Techniques for Hazardous Polluting Substances, USEPA.

Maryland Department of Health and Mental Hygiene, Maryland Code of Regulations Title 10, Subtitle 51, Disposal of Designated Hazardous Substances.

__________, Subtitle 18, Air Pollution Regulations.

Maryland Department of Natural Resources, Code of Maryland Regulations - 08:05:04:07 - Prevention of Oil Pollution.

Maryland State Fire Commission, Code of Maryland Regulations, Title 12, Subtitle 03, Chapter 01, Fire Prevention Code.

McKinnon, G., ed., 1979, Industrial Fire Hazards Handbook, National Fire Protection Association, Boston, MA.

National Fire Protection Association, 1978, Fire Protection Guide on Hazardous Materials, National Fire Protection Association, Boston, MA.

________, 1982, National Fire Codes 30, 43A, 77, and 329, National Fire Protection Association, Boston, MA.

Ontario Ministry of the Environment, Waste Management Branch, 1978, Guidelines for Environment Protection Measures at Chemical Storage Facilities.

Petroleum Association for Conservation of the Canadian Environment, 1980, Bulk Plant Guidelines for Oil Spill Prevention and Control, Pace Report No. 80-3.

Scott, R., 1979, Toxic Chemical and Explosives Facilities: Safety and Engineering Design, 1979, ASC Symposium 96, American Chemical Society, Washington, D.C.

Sittig, Marshall, 1974, Oil Spill Prevention and Removal Handbook, Noyes Data Corportion, Park Ridge, NJ.

Texas A & M Research Foundation, 1978, Oil Spill Control Course Handbook, Texas Engineering Extension Service, College Station, TX.

D.J. DeRenzo, ed., 1978, Unit Operations for Treatment of Hazardous Industrial Wastes, Noyes Data Corporation, Park Ridge, NJ.

United States Environmental Protection Agency, 1980, Standards Applicable to Owners and Operators of Hazardous Waste Treatment, Storage and Disposal Facilities Under RCRA, Subtitle C, Section 3004; General and Interim Standards for Tanks; Interim Status Standards for Chemical, Physical, and Biological Treatment, Washington, D.C.

________, Oil Pollution Prevention, 40 CFR Part 112, Washington, D.C.

Appendix A
List of Chemical Representatives by Class

Class 1 Acids, Mineral, Non-Oxidizing

Boric acid*
Chlorosulfonic acid*
Difluorophosphoric acid
Disulfuric acid
Fluoroboric acid
Fluorosulfonic acid
Fluosilicic acid
Hexafluorophosphoric acid
Hydriodic acid*
Hydrobromic acid*
Hydrochloric acid*
Hydrocyanic acid*
Hydrofluoric acid*
Monofluorophosphoric acid
Permonosulfuric acid
Phosphoric acid*
Selenous acid

Class 2 Acids, Mineral, Oxidizing

Bromic acid
Chloric acid*
Chromic acid*
Hypochlorous acid
Nitric acid*
Nitrohydrochloric acid
Oleum*
Perbromic acid
Perchloric acid*
Perchlorous acid
Periodic acid
Sulfuric acid*
Sulfur trioxide*

Class 3 Acids, Organic (All Isomers)

Acetic acid*
Acrylic acid
Adipic acid
Benzoic acid*
Butyric acid
Capric acid
Caproic acid
Caprylic acid
Chloromethylphenoxyacetic acid
Cyanoacetic acid
Dichlorophenoxyacetic acid
Endothal
Fluoracetic acid
Formic acid*
Fumaric acid
Glycolic acid
Hydroxydibromobenzoic acid
Lactic acid*
Maleic acid*
Monochloroacetic acid
Oleic acid*
Oxalic acid
Peracetic acid
Phenoxyacetic acid*
Phthalic acid*
Propionic acid
Salycilic acid*
Succinic acid
Trichlorophenoxyacetic acid
Trinitrobenzoic acid
Toluic acid
Valeric acid

Class 4 Alcohols and Glycols (All Isomers)

Acetone cyanohydrin
Allyl alcohol*
Aminoethanol
Amyl alcohol
Benzyl alcohol
Butanediol
Butyl alcohol
Butyl cellosolve*
Chloroethanol*
Crotyl alcohol
Cyclohexanol*
Cyclopentanol
Decanol
Diacetone alcohol
Dichloropropanol
Diethanol amine
Diisopropanolamine
Ethanol*
Ethoxyethanol
Ethylene chlorohydrin*
Ethylene cyanohydrin
Ethylene glycol*
Ethylene glycol monomethyl ether*
Glycerin*
Heptanol
Hexanol
Isobutanol
Isopropanol
Mercaptoethanol
Methanol*
Monoethanol amine*
Monoisopropanol amine
Nonanol
Octanol
Propanol
Propylene glycol

*Representative chemical found in compatibility matrices.

Propylene glycol monomethyl ether
Triethanolamine

Class 5 Aldehydes (All Isomers)

Acetaldehyde*
Acrolein*
Benzaldehyde
Butyraldehyde
Chloral hydrate
Chloracetaldehyde
Crotonaldehyde
Formaldehyde*
Furfural*
Glutaraldehyde
Heptanal
Hexanal
Nonanal
Octanal
Propionaldehyde
Tolualdehyde
Urea formaldehyde
Valeraldehyde

Class 6 Amides (All Isomers)

Acetamide*
Benzadox
Bromobenzoyl acetanilide
Butyramide
Carbetamide
Diethylamide*
Diethyltoluamide
Dimethylformamide*
Dimefox
Diphenamide
Fluroacetanilide
Formamide
Propionamide
Schradan
Tris-(1-aziridinyl) phosphine oxide
Wepsyn* 155
Valeramide

Class 7 Amines, Aliphatic and Aromatic (All Isomers)

Aminodiphenyl
Aminoethanol*
Aminoethanolamine
Aminophenol
Aminopropionitrile
Amylamine
Aminothiazole
Aniline*
Benzidine
Benzylamine
Butylamine
Chlorotoluidine
Crimidine
Cupriethylenediamine
Cyclohexylamine
Diamine*
Dichlorobenzidine
Diethanolamine
Diethylamine*
Diethylenetriamine
Diisopropanolamine
Dimethylamine
Dimethylaminoazobenzene
Diphenylamine
Diphenylamine chloroarsine
Dipicrylamine
Dipropylamine
Ethylamine
Ethylenendiamine*
Ehtyleneimine
Hexamethylenediamine
Hexamethylenetetraamine
Hexylamine
Isopropylamine
Methylamine*
N-Methyl aniline
4,4-Methylene bis(2-chloroaniline)
Methyl ethyl pyridine
Monoethanolamine*
Monoisopropanolamine
Morpholine
Naphthylamine
Nitroaniline*
Nitrogen mustard
Nitrosodimethylamine
Pentylamine
Phenylene diamine
Picramide
Picridine
Piperidine
Propylamine
Propyleneimine
Pyridine*
Tetramethylenediamine
Toluidine
Triethanolamine
Triethylamine
Triethylenetetraamine
Trimethylamine
Tripropylamine

Class 8 Azo Compounds, Diazo Compounds, and Hydrazines (All Isomers)

Aluminum tetraazidoborate
Aminothiazole
Azidocarbonyl guanidine
Azido-s-triazole
a.á-Azodiisobutyronitrile
Benzene diazonium chloride
Benzotriazole
t-Butyl azidoformate
Chloroazodin
Chlorobenzotriazole
Diazodinitrophenol
Diazidoethane
Dimethylamino azobenzene
Dimethyl hydrazine*
Dinitrophenyl hydrazine
Guanyl nitrosoaminoguanylidine hydrazine
Hydrazine*
Hydrazine azide
Methyl hydrazine
Mercaptobenzothiazole
Phenyl hydrazine hydrochloride
Tetrazene

Class 9 Carbamates

Aldicarb
Bassa*
Baygon*
Butacarb
Bux*
Carbaryl
Carbanolate
Dioxacarb
Dowco* 139
Formetanate hydrochloride
Furadan*
Hopcide*
N-Isopropylmethylcarbamate
Landrin*
Matacil*
Meobal
Mesurol*
Methomyl
Mipcin*
Mobam*
Oxamyl
Pirimicarb
Promecarb
Tranid*
Tsumacide*

Class 10 Caustics

Ammonia*
Ammonium hydroxide*
Barium hydroxide
Barium oxide
Beryllium hydroxide
Cadmium amide
Calcium hydroxide*
Calcium oxide*
Lithium amide
Lithium hydroxide
Potassium aluminate
Potassium butoxide
Potassium hydroxide
Sodium aluminate
Sodium amide
Sodium carbonate*
Sodium hydroxide*
Sodium hypochlorite
Sodium methylate
Sodium oxide

Class 11 Cyanides

Cadmium cyanide
Copper cyanide
Cyanogen bromide
Hydrocyanic acid*
Lead cyanide
Mercuric cyanide
Mercuric oxycyanide
Nickel cyanide
Potassium cyanide*
Silver cyanide
Sodium cyanide*
Zinc cyanide

Class 12 Dithiocarbamates

CDEC
Dithane* M-45
Ferbam
Maneb
Metham
Nabam
Niacide*
Polyram-cobi*
Selenium diethyl dithiocarbamate
Thiram
Zinc salts of dimethyl dithiocarbamic acid
Zineb
Ziram

Group 13 Esters (All Isomers)

Allyl chlorocarbonate
Amyl acetate
Butyl acetate*
Butyl acrylate

Butyl benzyl phthalate
Butyl formate
Dibutyl phthalate
Diethylene glycol monobutyl ether acetate
Ethyl acetate*
Ethyl butyrate
Ethyl chloroformate
Ethyl formate
2-Ethyl hexylacrylate
Ethyl propionate
Glycol diacetate
Isobutyl acetate
Isobutyl acrylate
Isodecyl acrylate
Isopropyl acetate
Medinoterb acetate
Methyl acetate
Methyl acrylate
Methyl amyl acetate
Methyl butyrate
Methyl chloroformate
Methyl formate*
Methyl methacrylate
Methyl proprionate
Methyl valerate
Propiolactone*
Propyl acetate
Propyl formate
Vinyl acetate
Dimethyl phthalate*

Class 14 Ethers (All Isomers)

Anisole
Butyl cellosolve*
Bromodimethoxyaniline
Dibutyl ether
Dichloroethyl ether*
Diethyl ether*
Dimethyl ether
Dimethyl formal
Dioxane*
Diphenyl oxide
Ethoxyethanol
Ethylene glycol monomethyl ether*
Furan*
Glycol ether
Isopropyl ether
Methyl butyl ether
Methyl chloromethyl ether
Methyl ethyl ether
Polyglycol ether
Propyl ether
Propylene glycol monomethyl ether
TCDD
Tetrachloropropyl ether
Tetrahydrofuran*
Trinitroanisole
Vinyl ethyl ether
Vinyl isopropyl ether

Class 15 Fluorides, Inorganic

Aluminum fluoride*
Ammonium bifluoride
Ammonium fluoride*
Barium fluoride
Beryllium fluoride
Cadmium fluoride
Calcium fluoride
Cesium fluoride
Chromic fluoride
Fluoroboric acid
Fluosilicic acid*
Fluorosilicic acid*
Hexafluorophosphoric acid
Hydrofluoric acid*
Hydrofluorosilicic acid*
Magnesium fluoride
Potassium fluoride
Selenium fluoride
Silicon tetrafluoride
Sodium fluoride
Sulfur pentafluoride
Tellurium hexafluoride
Zinc fluoroborate

Class 16 Hydrocarbons, Aromatic (All Isomers)

Acenaphthene
Anthracene
Benz-a-pyrene
Benzene*
n-Butyl benzene
Chrysene
Cumene*
Cymene
Decyl benzene
Diethyl benzene
Diphenyl
Diphenyl acetylene
Diphenyl ethane

Diphenyl ethylene
Diphenyl methane
Dodecyl benzene
Dowtherm
Durene
Ethyl benzene*
Fluoranthrene
Fluorene
Hemimellitene
Hexamethyl benzene
Indene
Isodurene
Mesitylene
Methyl naphthalene
Naphthalene*
Pentamethyl benzene
Phenanthrene
Phenyl acetylene
Propyl benzene
Pseudocumene
Styrene*
Tetraphenyl ethylene
Toluene*
Stilbene
Triphenylethylene
Triphenylmethane
Xylene*

Class 17 Halogenated Organics (All Isomers)

Acetyl bromide
Acetyl chloride
Aldrin*
Allyl bromide
Allyl chloride
Allyl chlorocarbonate
Amyl chloride
Benzal bromide
Benzal chloride
Benzotribromide
Benzotrichloride
Benzyl bromide
Benzyl chloride*
Benzyl chlorocarbonate
Bromoacetylene
Bromobenzyl trifluoride
Bromoform
Bromophenol
Bromopropyne
Bromotrichloromethane
Bromotrifluoromethane
Bromoxynil
Butyl fluoride
Carbon tetrachloride*
Carbon tetrafluoride
Carbon tetraiodide
Chloral hydrate
Chlordane
Chloracetaldehyde
Chloroacetic acid
Chloroacetone*
Chloroacetophenone
Chloroacrylonitrile
Chloranil (tetrachloroquinone)
Chloroazodin
Chlorobenzene*
Chlorobenzotriazole
Chlorobenzoyl peroxide
Chlorobenzylidene malononitrile
Chlorobutyronitrile
Chlorocresol*
Chlorodinitrotoluene
Chloroethanol*
Chloroethylenimime
Chloroform*
Chlorohydrin
Chloromethyl methyl ether
Chloromethyl phenoxyacetic acid
Chloronitroaniline
Chlorophenol
Chlorophenyl isocyanate
Chloropicrin*
Chlorothion
Chlorotoluidine
CMME
Crotyl bromide
Crotyl chloride (1-chloro-2-butene)
DDD
DDT
DDVP
Dibromochloropropane
Dichloroacetone*
Dichlorobenzene
Dichlorobenzidine
Dichloroethane
Dichloroethylene
Dichloroethyl ether*
Dichloromethane (methylene dichloride)*
Dichlorophenol
Dichlorophenoxy acetic acid
Dichloropropane
Dichloropropanol
Dichloropropylene
Dieldrin
Diethyl chloro vinyl phosphate
Dichlorophene
Dinitrochlorobenzene

Endosulfan
Endrin
Epichlorohydrin*
Ethyl chloroformate
Ethylene chlorohydrin*
Ethylene dibromide
Ethylene dichloride*
Fluoroacetanilide
Freons*
Heptachlor
Hexachlorobenzene
Hydroxydibromobenzoic acid
Isopropyl chloride
α-Isopropyl methyl phosphoryl fluoride
Lindane
Methyl bromide
Methylchloride*
Methyl chloroform
Methyl chloroformate
Methyl ethyl chloride
Methyl iodide
Monochloroacetone
Nitrochlorobenzene
Nitrogen mustard
Pentachlorophenol*
Perchloroethylene
Perchloromethylmercaptan
Picryl chloride
Polybrominated biphenyls
Polychlorinated biphenyls
Polychlorinated triphenyls
Propargyl bromide
Propargyl chloride
TCDD
Tetrachloroethane*
Tetrachlorophenol
Tetrachloropropyl ether
Trichloroethane
Trichloroethylene*
Trichlorophenoxyacetic acid
Trichloropropane
Trifluoroethane
Vinyl chloride
Vinylidene chloride (1,1-dichloroethylene)

Class 18 Isocyanates (All Isomers)

Chlorophenyl isocyanate
Diphenylmethane diisocyanate
Methyl isocyanate
Methylene diisocyanate
Polyphenyl polymethylisocyanate
Toluene diisocyanate*

Class 19 Ketones (All Isomers)

Acetone*
Acetophenone*
Acetyl acetone*
Benzophenone
Bromobenzoyl acetanilide
Chloroacetophenone
Coumafuryl
Coumatetralyl
Cyclohexanone*
Diacetone alcohol
Diacetyl
Dichloroacetone*
Diethyl ketone
Dimethyl ketone*
Diisobutyl ketone
Heptanone
Hydroxyacetophenone
Isophorone
Mesityl oxide
Methyl t-butyl ketone
Methyl ethyl ketone*
Methyl isobutyl ketone*
Methyl isopropenyl ketone
Methyl n-propyl ketone
Methyl vinyl ketone
Monochloroacetone
Nonanone
Octanone
Pentanone
Quinone (Benzoquinone)*

Class 20 Mercaptans and Other Organic Sulfides (All Isomers)

Aldicarb
Amyl mercaptan
Butyl mercaptan
Carbon disulfide*
Dimethyl sulfide
Endosulfan
Ethyl mercaptan*
Mercaptobenzothiazole
Mercaptoethanol
Methomyl
Methyl mercaptan

Naphthyl mercaptan
Perchloromethyl mercaptan
Phospholan
Polysulfide polymer
Propyl mercaptan
Sulfur mustard
Tetrasul
Thionazin

Class 21 Metal Compounds, Inorganic

Aluminum fluoride*
Aluminum sulfate*
Ammonium arsenate
Ammonium dichromate
Ammonium hexanitrocobaltate
Ammonium molybdate
Ammonium nitridoosmate
Ammonium permanganate
Ammonium tetrachromate
Ammonium tetraperoxychromate
Ammonium trichromate
Antimony
Antimony nitride
Antimony oxychloride
Antimony pentachloride
Antimony pentafluoride
Antimony pentasulfide
Antimony perchlorate
Antimony potassium tartrate
Antimony sulfate
Antimony tribromide
Antimony trichloride
Antimony triiodide
Antimony trifluoride
Antimony trioxide
Antimony trisulfide
Antimony trivinyl
Arsenic
Arsenic pentaselenide
Arsenic pentoxide
Arsenic pentasulfide
Arsenic sulfide
Arsenic tribromide
Arsenic trichloride
Arsenic trifluoride
Arsenic triiodide
Arsenic trisulfide
Arsines
Barium
Barium azide
Barium carbide
Barium chlorate
Barium chloride
Barium chromate
Barium fluoride
Barium fluosilicate
Barium hydride
Barium hydroxide
Barium hypophosphide
Barium iodate
Barium iodide
Barium nitrate
Barium oxide
Barium perchlorate
Barium permanganate
Barium peroxide
Barium phosphate
Barium stearate
Barium sulfide
Barium sulfite
Beryllium
Beryllium-copper alloy
Beryllium fluoride
Beryllium hydride
Beryllium hydroxide
Beryllium oxide
Beryllium tetradhydroborate
Bismuth
Bismuth chromate
Bismuthic acid
Bismuth nitride
Bismuth pentafluoride
Bismuth pentoxide
Bismuth sulfide
Bismuth tribromide
Bismuth trichloride
Bismuth triiodide
Bismuth trioxide
Borane
Bordeaux arsenites
Boron arsenotribromide
Boron bromodiodide
Boron dibromoiodide
Boron nitride
Boron phosphide
Boron triazide
Boron tribromide
Boron triiodide
Born trisulfide
Boron trichloride
Boron trifluoride
Cacodylic acid
Cadmium
Cadmium acetylide

Cadmium amide
Cadmium azide
Cadmium bromide
Cadmium chlorate
Cadmium chloride
Cadmium cyanide
Cadmium fluoride
Cadmium hexamine chlorate
Cadmium hexamine perchlorate
Cadmium iodide
Cadmium nitrate
Cadmium nitride
Cadmium oxide
Cadmium phosphate
Cadmium sulfide
Cadmium trihydrazine chlorate
Cadmium trihydrazine perchlorate
Calcium arsenate
Calcium arsenite
Chromic acid*
Chromic chloride
Chromic fluoride
Chromic oxide
Chromic sulfate
Chromium
Chromium sulfide
Chromium trioxide
Chromyl chloride
Cobalt
Cobaltous bromide
Cobaltous chloride
Cobaltous nitrate
Cobaltous sulfate
Cobaltous resinate
Copper
Copper acetoarsenite
Copper acetylide
Copper arsenate
Copper arsenite
Copper chloride
Copper chlorotetrazole
Copper cyanide
Copper nitrate
Copper nitride
Copper sulfate
Copper sulfide
Cupriethylene diamine
Cyanochloropentane
Diethyl zinc
Diisopropyl beryllium
Diphenylamine chloroarsine
Ethyl dichloroarsine
Ethylene chromic oxide
Ferric arsenate
Ferrous arsenate
Hydrogen selenide
Indium
Lead
Lead acetate
Lead arsenate
Lead arsenite
Lead azide
Lead carbonate
Lead chlorite
Lead cyanide
Lead dinitroresordinate
Lead monoinitroresorcinate
Lead nitrate
Lead oxide
Lead styphnate
Lead sulfide
Lewisite
London purple
Magnesium arsenate
Magnesium arsenite
Manganese
Manganese acetate
Manganese arsenate
Manganese bromide
Manganese chloride
Manganese methylcyclopentadienyl tricarbonyl
Manganese nitrate
Manganese sulfide
Mercuric acetate
Mercuric ammonium chloride
Mercuric benzoate
Mercuric bromide
Mercuric chloride
Mercuric cyanide
Mercuric iodide
Mercuric nitrate
Mercuric oleate
Mercuric oxide
Mercuric oxycyanide
Mercuric potassium iodide
Mercuric salicylate
Mercuric subsulfate
Mercuric sulfate
Mercuric sulfide
Mercuric thiocyanide
Mercurol
Mercurous bromide
Mercurous gluconate
Mercurous iodide

Mercurous nitrate
Mercurous oxide
Mercurous sulfate
Mercury
Mercury fulminate
Methoxyethylmercuric chloride
Methyl dichloroarsine
Molybdenum
Molybdenum sulfide
Molybdenum trioxide
Molybdic acid
Nickel
Nickel acetate
Nickel antimonide
Nickel arsenate
Nickel arsenite
Nickel carbonyl
Nickel chloride
Nickel cyanide
Nickel nitrate
Nickel selenide
Nickel subsulfide
Nickel sulfate
Osmium
Osmium amine nitrate
Osmium amine perchlorate
Phenyl dichloroarsine
Potassium arsenate
Potassium arsenite
Potassium dichromate
Potassium permanganate
Selenium
Selenium fluoride
Selenium diethyl dithiocarbamate
Selenous acid
Silver acetylide
Silver azide
Silver cyanide
Silver nitrate*
Silver nitride
Silver styphnate
Silver sulfide
Silver tetrazene
Sodium arsenate
Sodium arsenite
Sodium cacodylate
Sodium chromate
Sodium dichromate
Sodium molybdate
Sodium permanganate
Sodium selenate
Stannic chloride
Stannic sulfide
Strontium arsenate
Strontium monosulfide
Strontium nitrate
Strontium peroxide
Strontium tetrasulfide
Tellurium hexafluoride
Tetraethyl lead*
Tetramethyl lead
Tetraselenium tetranitride
Thallium
Thallium nitride
Thallium sulfide
Thallous sulfate
Thorium
Titanium
Titanium sulfate
Titanium sesquisulfide
Titanium tetrachloride
Titanium sulfide
Tricadmium dinitride
Tricesium nitride
Triethyl arsine
Triethyl bismuthine
Triethyl stibine
Trilead dinitride
Trimercury dinitride
Trimethyl arsine
Trimethyl bismuthine
Trimethyl stibine
Tripropyl stibine
Trisilyl arsine
Trithorium tetranitride
Trivinyl stibine
Tungstic acid
Uranium sulfide
Uranyl nitrate
Vanadic acid anhydride
Vanadium oxytrichloride
Vanadium tetroxide
Vanadium trichloride
Vanadyl sulfate
Zinc
Zinc acetylide
Zinc ammonium nitrate
Zinc arsenate
Zinc arsenite
Zinc chloride*
Zinc cyanide
Zinc fluoborate

Zinc nitrate
Zinc permanganate
Zinc peroxide
Zinc phosphide
Zinc salts of dimethyldithio carbamic acid
Zinc sulfate
Zinc sulfide
Zirconium
Zirconium chloride
Zirconium picramate

Class 22 Nitrides

Antimony nitride
Bismuth nitride
Boron nitride
Copper nitride
Disulfur dinitride
Lithium nitride
Potassium nitride
Silver nitride
Sodium nitride
Tetraselenium tetranitride
Tetrasulfur tetranitride
Thallium nitride
Tricadmium dinitride
Tricalcium dinitride
Tricesium nitride
Trilead dinitride
Trimercury dinitride
Trithorium tetranitride

Class 23 Nitriles (All Isomers)

Acetone cyanohydrin
Acetonitrile*
Acrylonitrile*
Adiponitrile
Aminopropionitrile
Amyl cyanide
a,á-Azodiisobutyronitrile
Benzonitrile
Bromoxynil
Butyronitrile
Chloroacrylonitrile
Chlorobenzylidene malononitrile
Chlorobutyronitrile
Cyanoacetic acid
Cyanochloropentane
Cyanogen
Ethylene cyanohydrin
Glycolonitrile
Phenyl acetonitrile
Phenyl valerylnitrile
Propionitrile
Surecide*
Tetramethyl succinonitrile
Tranid*
Vinyl cyanide

Class 24 Nitro Compounds (All Isomers)

Acetyl nitrate
Chlorodinitroluene
Chloronitroaniline
Chloropicrin
Collodion
Diazodinitrophenol
Diethylene glycol dinitrate
Dinitrobenzene
Dinitrochlorobenzene
Dinitrocresol
Dinitrophenol
Dinitrophenyl hydrazine
Dinitrotoluene
Dinoseb
Dipentaerythritol hexanitrate
Dipicryl amine
Ethyl nitrate
Ethyl nitrite
Glycol dinitrate
Glycol monolactate trinitrate
Guanidine nitrate
Lead dinitroresorcinate
Lead mononitroresorcinate
Lead styphnate
Mannitol hexanitrate
Medinoterb acetate
Nitroaniline*
Nitrobenzene*
Nitrobiphenyl
Nitrocellulose
Nitrochlorobenzene
Nitroglycerin
Nitrophenol*
Nitropropane*
N-Nitrosodimethylamine
Nitrosoguanidine
Nitrostarch
Nitrotoluene*
Nitroxylene
Pentaerythritol tetranitrate
Picramide
Picric acid*
Picryl chloride

Polyvinyl nitrate
Potassium dinitrobenzfuroxan
RDX
Silver styphnate
Sodium picramate
Tetranitromethane
Trinitroanisole
Trinitrobenzene
Trinitrobenzoic acid
Trinitronaphthalene
Trinitroresorcinol
Trinitrotoluene
Urea nitrate

Class 25 Hydrocarbons, Aliphatic, Unsaturated (All Isomers)

Acetylene
Allene
Amylene
Butadiene*
Butene
Cyclopentene
Decene
Dicyclopentadiene
Diisobutylene
Dimethyl acetylene
Dimethyl butyne
Dipentene
Dodecene
Ethyl acetylene
Ethylene
Heptene
Hexene
Hexyne
Isobutylene
Isooctene
Isoprene*
Isopropyl acetylene
Methyl acetylene
Methyl butene
Methyl butyne
Methyl styrene
Nonene
Octadecyne
Octene
Pentene
Pentyne
Polybutene
Polypropylene
Propylene
Styrene*
Tetradecene
Tridecene
Undecene
Vinyl toluene

Class 26 Hydrocarbons, Aliphatic, Saturated

Butane*
Cycloheptane
Cyclohexane*
Cyclopentane
Cyclopropane
Decalin
Decane
Ethane
Heptane
Hexane
Isobutane
Isohexane
Isooctane
Isopentane
Methane
Methyl cyclohexane
Neohexane
Nonane
Octane
Pentane
Propane

Class 27 Peroxides and Hydroperoxides, Organic (All Isomers)

Acetyl benzoyl peroxide
Acetyl peroxide
Benzoyl peroxide*
Butyl hydroperoxide
Butyl peroxide
Butyl peroxyacetate
Butyl peroxybenzoate
Butyl peroxypivalate
Caprylyl peroxide
Chlorobenzoyl peroxide
Cumene hydroperoxide
Cyclohexanone peroxide
Dicumyl peroxide
Diisopropylbenzene hydroperoxide
Diisopropyl peroxydicarbonate
Dimethylhexane dihydroperoxide
Hydrogen peroxide*
Isopropyl percarbonate
Lauroyl peroxide
Methyl ethyl ketone peroxide
Peracetic acid
Succinic acid peroxide

Class 28 Phenols, Cresols (All Isomers)

Amino phenol
Bromophenol
Bromoxynil
Carbacrol
Carbolic oil
Catecol
Chlorocresol*
Chlorophenol
Coal tar*
Cresol*
Creosote*
Cyclohexyl phenol
Dichlorophenol
Dinitrocresol
Dinitrophenol
Dinoseb
Eugenol
Guaiacol
Hydroquinone*
Hydroxyacetophenone
Hydroxydiphenol
Hydroxyhydroquinone
Isoeugenol
Naphthol
Nitrophenol*
Nonyl phenol
Pentachlorophenol
Phenol*
o-Phenyl phenol
Phloroglucinol
Picric acid*
Pyrogallol
Resorcinol*
Saligenin
Sodium pentachlorophenate
Sodium phenolsulfonate
Tetrachlorophenol
Thymol*
Trichlorophenol
Trinitroresorcinol

Class 29 Organophosphates, Phosphothioates, and Phosphodithioates

Abate*
Azinphos ethyl
Azodrin*
Bidrin*
Bomyl*
Chlorfenvinphos
Chlorothion*
Coroxon*
DDVP
Demeton
Demeton-s-methyl sulfoxid
Diazinon*
Diethyl chlorovinyl phosphate
Dimethyldithiophosphoric acid
Dimefox
Dioxathion
Disulfoton
Dyfonate*
Endothion
EPN
Ethion*
Fensulfothion
Guthion*
Hexaethyl tetraphosphate
Malathion*
Mecarbam
Methyl parathion
Mevinphos
Mocap*
α-Isopropyl methylphosphoryl fluoride
Paraoxon
Parathion*
Phorate
Phosphamidon
Phospholan
Potasan
Prothoate
Shradan
Sulfotepp
Supracide*
Shradan
Sulfotepp
Supracide*
Surecide*
Tetraethyl dithionopyrophosphate
Tetraethyl pyrophosphate
Thionazin
Tris-(1-aziridinyl) phosphine oxide
VX
Wepsyn* 155

Class 30 Sulfides, Inorganic

Ammonium sulfide
Antimony pentasulfide
Antimony trisulfide
Arsenic pentasulfide
Arsenic sulfide
Arsenic trisulfide
Barium sulfide
Beryllium sulfide
Bismuth sulfide

Bismuth trisulfide
Boron trisulfide
Cadmium sulfide
Calcium sulfide
Cerium trisulfide
Cesium sulfide
Chromium sulfide
Copper sulfide
Ferric sulfide
Ferrous sulfide
Germanium sulfide
Gold sulfide
Hydrogen sulfide
Lead sulfide
Lithium sulfide
Magnesium sulfide
Manganese sulfide
Mercuric sulfide
Molybdenum sulfide
Nickel subsulfide
Phosphorous heptasulfide
Phosphorous pentasulfide
Phosphorous sesquisulfide
Phosphorous trisulfide
Potassium sulfide
Silver sulfide
Sodium sulfide
Stannic sulfide
Strontium monosulfide
Strontium tetrasulfide
Thallium sulfide
Titanium sesquisulfide
Titanium sulfide
Uranium sulfide
Zinc sulfide

Class 31 Epoxides

Butyl glycidyl ether
t-Butyl-3-phenyl oxazirane
Cresol glycidyl ether
Diglycidyl ether
Epichlorohydrin*
Epoxybutane
Epoxybutene
Epoxyethylbenzene
Ethylene oxide
Glycidol
Phenyl glycidyl ether
Propylene oxide

Class 32 Combustible and Flammable Materials, Miscellaneous

Alkyl resins
Asphalt
Bakelite*
Buna-N*
Bunker fule oil
Camphor oil
Carbon, activated, spent
Cellulose
Coal oil
Diesel oil*
Dynes thinner
Gas oil, cracked
Gasoline*
Grease
Isotactic propylene
J-100
Jet oil
Kerosene*
Lacquer thinner
Methyl acetone
Mineral spirits
Naphtha*
Oil of bergamot
Orris root
Paper
Petroleum naphtha
Petroleum oil*
Polyamide resin
Polyester resin
Polyethylene
Polymeric oil
Polypropylene
Polystyrene
Polysulfide polymer
Polyurethane
Polyvinyl acetate
Polyvinyl chloride
Refuse
Resins
Sodium polysulfide
Stoddard solvent
Sulfur (elemental)
Synthetic rubber
Tall oil
Tallow
Tar
Turpentine*
Unisolve
Waxes
Wood

Class 33 Explosives

Acetyl azide
Acetyl nitrate
Ammonium azide
Ammonium chlorate
Ammonium hexanitrocobaltate
Ammonium nitrate
Ammonium nitrite
Ammonium periodate
Ammonium permanganate
Ammonium picrate
Ammonium tetraperoxychromate
Azidocarbonyl guanidine
Barium azide
Benzene diazonium chloride
Benzotriazole
Benzoyl peroxide*
Bismuth nitride
Boron triazide
Bromine azide
Butanetriol trinitrate
t-Butyl hypochlorite
Cadmium azide
Cadmium haxamine chlorate
Cadmium hexamine perchlorate
Cadmium nitrate
Cadmium nitride
Cadmium trihydrazine chlorate
Calcium nitrate
Cesium azide
Chlorine azide
Chlorine dioxide
Chlorine fluoroxide
Chlorine trioxide
Chloroacetylene
Chloropicrin
Copper acetylide
Cyanuric triazide
Diazidoethane
Diazodinitrophenol
Diethylene glycol dinitrate
Dipentaerithritol hexanitrate
Dipicryl amine
Disulfur dinitride
Ethyl nitrate
Ethyl nitrite
Fluorine azide
Glycol dinitrate
Glycol monolactate trinitrate
Gold fulminate
Guanyl nitrosaminoguanylidene hydrazine
HMX
Hydrazine azide
Hydrazoic acid
Lead azide
Lead dinitroresorcinate
Lead mononitroresorcinate
Lead styphnate
Mannitol hexanitrate
Mercuric oxycyanide
Mercury fulminate
Nitrocarbonitrate
Nitrocellulose
Nitroglycerin
Nitrosoguanidine
Nitrostarch
Pentaerythritol tetranitrate
Picramide
Picric acid*
Picryl chloride
Polyvinyl nitrate
Potassium dinitrobenzfuroxan
Potassium nitrate
RDX
Silver acetylide
Silver azide
Silver nitride
Silver styphnate
Silver tetrazene
Smokeless powder
Sodium azide
Sodium picramate
Tetranitromethane
Tetraselenium tetranitride
Tetrasulfur tetranitride
Tetrazene
Thallium nitride
Trilead dinitride
Trimercury dinitride
Trinitrobenzene
Trinitrobenzoic acid
Trinitronaphthalene
Trinitroresorcinol
Trinitrotoluene
Urea nitrate
Vinyl azide
Zinc peroxide

Class 34 Polymerizable Compounds

Acrolein
Acrylic acid
Acrylonitrile*
Butadiene*
n-Butyl acrylate
Ethyl acrylate
Ethylene oxide
Ethylenimine
2-Ethylhexyl acrylate
Isobutyl acrylate
Isoprene

Methyl acrylate*
Methyl methacrylate
2-Methyl styrene
Propylene oxide
Styrene*
Vinyl acetate
Vinyl chloride
Viyl cyanide
Vinylidene chloride
Vinyl toluene

<u>Class 35 Oxidizing Agents, Strong</u>

Ammonium chlorate
Ammonium dichromate
Ammonium nitridoosmate
Ammonium perchlorate
Ammonium periodate
Ammonium permanganate
Ammonium persulfate
Ammonium tetrachromate
Ammonium tetraperoxychromate
Ammonium trichromate
Antimony perchlorate
Barium bromate
Barium chlorate
Barium iodate
Barium nitrate
Barium perchlorate
Barium permanganate
Barium peroxide
Bromic acid
Bromine
Bromine monofluoride
Bromine pentafluoride
Bromine trifluoride
t-Butyl hypochlorite
Cadmium chlorate
Cadmium nitrate
Calcium bromate
Calcium chlorate
Calcium chlorite
Calcium hypochlorite
Calcium iodate
Calcium nitrate
Calcium perchromate
Calcium permanganate
Calcium peroxide
Chloric acid*
Chlorine
Chlorine dioxide
Chlorine fluoroxide
Chlorine monofluoride
Chlorine monoxide
Chlorine pentafluoride
Chlorine trifluoride
Chlorine trioxide
Chromic acid*
Chromyl chloride
Cobaltous nitrate
Copper nitrate
Dichloroamine
Dichloroisocyanuric acid
Ethylene chromic oxide
Fluorine
Fluorine monoxide
Guanidine nitrate
Hydrogen peroxide
Iodine pentoxide
Lead chlorite
Lead nitrate
Lithium hypochlorite
Lithium peroxide
Magnesium chlorate
Magnesium nitrate
Magnesium perchlorate
Magnesium peroxide
Manganese nitrate
Mercuric nitrate
Mercurous nitrate
Nickel nitrate
Nitrogen dioxide
Osmium amine nitrate
Osmium amine perchlorate
Oxygen difluoride
Perchloryl fluoride
Phosphorus oxybromide
Phosphorus oxychloride
Potassium bromate
Potassium dichloroisocyanurate
Potassium dichromate
Potassium nitrate
Potassium perchlorate
Potassium permanganate
Potassium peroxide
Silver nitrate*
Sodium bromate
Sodium carbonate peroxide
Sodium chlorate
Sodium chlorite
Sodium dichloroisocyanurate
Sodium dichromate
Sodium hypochlorite*
Sodium nitrate
Sodium nitrite
Sodium perchlorate
Sodium permanganate
Sodium peroxide
Strontium nitrate
Strontium peroxide
Sulfur trioxide*

Trichloroisocyanuric acid
Uranyl nitrate
Urea nitrate
Zinc ammonium nitrate
Zinc nitrate
Zinc permanganate
Zinc peroxide
Zirconium picramate

Class 36 Reducing Agents, Strong

Aluminum borohydride
Aluminum carbide
Aluminum hydride
Aluminum hypophosphide
Ammonium hypophosphide
Ammonium sulfide
Antimony pentasulfide
Antimony trisulfide
Arsenic sulfide
Arsenic trisulfide
Arsine
Barium carbide
Barium hydride
Barium hypophosphide
Barium sulfide
Benzyl silane
Benzyl sodium
Beryllium hydride
Beryllium sulfide
Beryllium tetrahydroborate
Bismuth sulfide
Boron arsenotribromide
Boron trisulfide
Bromodiborane
Bromosilane
Butyl dichloroborane
n-Butyl lithium
Cadmium acetylide
Cadmium sulfide
Calcium
Calcium carbide
Calcium hexammoniate
Calcium hydride
Calcium hypophosphide
Calcium sulfide
Cerium hydride
Cerium trisulfide
Cerous phosphide
Cesium carbide
Cesium hexahydroaluminate
Cesium hydride
Cesium sulfide
Chlorodiborane
Chlorodiisobutyl aluminum
Chlorodimethylamie diborane
Chlorodipropyl borane
Chlorosilane
Chromium sulfide
Copper acetylide
Copper sulfide
Diamine*
Diborane
Diethyl aluminum chloride
Diethyl zinc
Diisopropyl beryllium
Dimethyl magnesium
Ferrous sulfide
Germanium sulfide
Gold acetylide
Gold sulfide
Hexaborane
Hydrazine*
Hydrogen selenide
Hydrogen sulfide
Hydroxyl amine
Lead sulfide
Lithium aluminum hydride
Lithium hydride
Lithium sulfide
Magnesium sulfide
Manganese sulfide
Mercuric sulfide
Methyl aluminum sesquibromide
Methyl aluminum sesquichloride
Methyl magnesium bromide
Methyl magnesium chloride
Methyl magnesium iodide
Molybdenum sulfide
Nickel subsulfide
Pentaborane
Phosphine
Phosphonium iodide
Phosphorus (red amorphous)
Phosphorus (white or yellow)
Phosphorus heptasulfide
Phosphorus pentasulfide
Phosphorus sesquisulfide
Phosphorus trisulfide
Potassium hydride
Potassium sulfide
Silver acetylide
Silver sulfide
Sodium
Sodium aluminate
Sodium aluminum hydride
Sodium hydride
Sodium hyposulfite
Sodium sulfide
Stannic sulfide
Strontium monosulfide
Strontium tetrasulfide

Tetraborane
Thallium sulfide
Titanium sesquisulfide
Titanium sulfide
Triethyl aluminum
Triethyl stibine
Triisobutyl aluminum
Trimethyl aluminum
Trimethyl stibine
Tri-n-butyl borane
Trioctyl aluminum
Uranium sulfide
Zinc acetylide
Zinc sulfide

Class 37 Water and Mixtures Containing Water

Aqueous solutions and mixtures
Water

Class 38 Water Reactive Substances

Acetic anhydride*
Acetyl bromide
Acetyl chloride
Alkyl aluminum chloride
Allyl tirchlorosilane
Aluminum aminoborohydride
Aluminum borohydride
Aluminum bromide
Aluminum chloride
Aluminum fluoride
Aluminum hydophosphide
Aluminum phosphide
Aluminum tetrahydroborate
Amyl trichlorosilane
Anisoyl chloride
Antimony tribromide
Antimony trichloride
Antimony trifluoride
Antimony triiodide
Antimony trivinyl
Arsenic tribromide
Arsenic trichloride
Arsenic triiodide
Barium
Barium carbide
Barium oxide
Barium sulfide
Benzene phosphorus dichloride
Benzoyl chloride
Benzyl silane
Benzyl sodium
Beryllium hydride
Beryllium tetrahydroborate
Bismuth pentafluoride
Borane
Boron bromodiiodide
Boron dibromoiodide
Boron phosphide
Boron tribromide
Boron trichloride
Boron trifluoride
Boron triiodide
Bromine monofluoride
Bromine pentafluoride
Bromine trifluoride
Bromo diethylaluminum
n-Butyl lithium
n-Butyl trichlorosilane
Cadmium acetylide
Cadmium amide
Calcium
Calcium carbide
Caldium hydride
Calcium oxide
Calcium phosphide
Cesium amide
Cesium hydride
Cesium phosphide
Chlorine dioxide
Chlorine monofluoride
Chlorine pentafluoride
Chlorine trifluoride
Chloroacetyl chloride
Chlorodiisobutyl aluminum
Chlorophenyl isocyanate
Chromyl chloride
Copper acetylide
Cyclohexenyl trichlorosilane
Cyclohexyl trichlorosilane
Decaborane
Diborane
Diethyl aluminum chloride
Diethyl dichlorosilane
Diethyl zinc
Diisopropyl beryllium
Dimethyl dichlorosilane
Dimethyl magnesium
Diphenyl dichlorosilane
Diphenylmethane diisocyanate
Disulfuryl chloride
Dodecyl trichlorosilane
Ethyl dichloroarsine
Ethyl dichlorosilane
Ethyl trichlorosilane
Fluorine
Fluorine monoxide

Fluorosulfonic acid
Gold acetylide
Hexadecyl trichlorosilane
Hexyl trichlorosilane
Hydrobromic acid*
Iodine monochloride
Lithium
Lithium aluminum hydride
Lithium amide
Lithium ferrosilicon
Lithium hydride
Lithium peroxide
Lithium silicon
Methyl aluminum sesquibromide
Methyl aluminum sesquichloride
Methyl dichlorosilane
Methylene diisocyanate
Methyl isocyanate
Methyl trichlorosilane
Methyl magnesium bromide
Methyl magnesium chloride
Methyl magnesium iodide
Nickel antimonide
Nonyl tirchlorosilane
Octadecyl trichlorosilane
Octyl trichlorosilane
Phenyl trichlorosilane
Phosphonium iodide
Phosphoric anhydride
Phosphorus oxychloride
Phosphorus pentasulfide
Phosphorus trisulfide
Phosphorus (amorphous red)
Phosphorus oxybromide
Phosphorus oxychloride
Phosphorus pentachloride
Phosphorus sesquisulfide
Phosphorus tribromide
Phosphorus trichloride
Polyphenyl polymethyl isocyanate
Potassium
Potassium hydride
Potassium oxide
Potassium peroxide
Propyl trichlorosilane
Pyrosulfuryl chloride
Silicon tetrachloride
Silver acetylide
Sodium
Sodium aluminum hydride
Sodium amide
Sodium hydride
Sodium methylate
Sodium oxide
Sodium peroxide
Sodium-potassium alloy
Stannic chloride
Sulfonyl fluoride
Sulfuric acid (70%)*
Sulfur chloride
Sulfur pentafluoride
Sulfur trioxide*
Sulfuryl chloride
Thiocarbonyl chloride
Thionyl chloride
Thiophosphoryl chloride
Titanium tetrachloride
Toluene diisocyanate
Trichlorosilane
Triethyl aluminum
Triisobutyl aluminum
Trimethyl aluminum
Tri-n-butyl aluminum
Tri-n-butyl borane
Trioctyl aluminum
Trichloroborane
Triethyl arsine
Triethyl stibine
Trimethyl arsine
Trimethyl stibine
Tripropyl stibine
Trisilyl arsine
Trivinyl stibine
Vanadium trichloride
Vinyl trichlorosilane
Zinc acetylide
Zinc phosphide
Zinc peroxide

Appendix B

Chemical Class Compatibility Matrix

Class Number	Chemical Class	1	2	3	4	5	6	7	8	9	10	11	12	13	14	15	16	17	18	19	20	21	22	23	24	25	26	27	28	29	30	31	32	33	34	35	36	37	38
1	Acids, Mineral, Non-oxidizing	1																																					
2	Acids, Mineral, Oxidizing		2																																				
3	Acids, Organic		G H	3																																			
4	Alcohols and Glycols	H	H F	H P	4																																		
5	Aldehydes	H P	H F	H P		5																																	
6	Amides	H	H GT				6																																
7	Amines, Aliphatic and Aromatic	H	H GT	H		H		7																															
8	Azo Compounds, Diazo Compounds, and Hydrazines	H G	H GT	H G	H G	H			8																														
9	Carbamates	H G	H GT						G H	9																													
10	Caustics	H	H	H		H				H G	10																												
11	Cyanides	GT GF	GT GF	GT GF					G			11																											
12	Dithiocarbamates	H GF F	H GF F	H GF GT		GF GT		U	H G				12																										
13	Esters	H	H F						H G		H			13																									
14	Ethers	H	H F												14																								
15	Fluorides, Inorganic	GT	GT	GT												15																							
16	Hydrocarbons, Aromatic		H F														16																						
17	Halogenated Organics	H GT	H F GT				H GT	H G		H GF	H							17																					
18	Isocyanates	H G	H F GT	H G	H P			H P	H G		H P G	H G	U						18																				
19	Ketones	H	H F						H G		H	H								19																			
20	Mercaptans and Other Organic Sulfies	GT GF	H F GT						H G									H	H	H	20																		
21	Metal Compounds, Inorganic	S	S	S			S	S			S											21																	
22	Nitrides	GF H F	H F E	H GF	GF H E	GF H			U	H G	U	GF H	GF H	GF H				GF H	U	GF H	GF H		22																
23	Nitriles	H GT GF	H F GT	H							U											S	GF H	23															
24	Nitro Compounds, Organic		H F GT			H					H E												H GE E		24														
25	Hydrocarbons, Aliphatic, Unsaturated	H	H F			H																				25													
26	Hydrocarbons, Aliphatic, Saturated		H F																								26												
27	Peroxides and Hydroperoxides, Organic	H G	H E		H F	H G		H GT	H F E	H F GT		H E GT	H F GT					H E	H	E	H F GT	H G	H GF E	H P WT		H P		27											
28	Phenols and Cresols	H	H F						H G										H P				GF H					H	28										
29	Organophosphates, Phosphothioates, Phosphodithioates	H GT	H GT						U		H E																	U		29									
30	Sulfides, Inorganic	GT GF	H F GT	GT		H			E										H									H GT			30								
31	Epoxides	H P	H P	H P	H P	U		H P	H P		H P	H P	U								H P	H P	H P					H P	H P	U	H P	31							
32	Combustible and Flammable Materials, Miscellaneous	H G	H F GT																				H GF F					H F GT					32						
33	Explosives	H E	H E	H E					H E		H E			H E								E	E					H E	H E		H E	H E	H E	33					
34	Polymerizable Compounds	P H	P H	P H					P H		P H	P H	U									P H	P H					P H	P H		P H			H E	34				
35	Oxidizing Agents, Strong	H GT		H GT	H F	H F	H F GT	H F GT	H E	H F GT		H E GT	H F GT	H F	H F		H F	H GT	H F GT	H F	H F GT		H F E	H F GT	H E	H F	H F	H G	H F	H F GT	H F GT	H F G	H F G	H E	H F GT	35			
36	Reducing Agents, Strong	H GF	H F GT	H GF	H GF F	GF H F	GF H	H GF	H G				H GT	H F				H E	GF H	GF H	GF H			H GF	H E			H E	GF H	GT GF H		H	GF H	H E	H P GF	H F E	36		
37	Water and Mixtures Containing Water	H	H						G										H G			S	GF H								GT GF						GF GT	37	
38	Water Reactive Substances	← EXTREMELY REACTIVE! DO NOT MIX WITH ANY CHEMICAL OR WASTE MATERIAL! EXTREMELY REACTIVE! →																																					38

LEGEND

Reactivity Code	Consequences
H	Heat generation
F	Fire
G	Innocuous and non-flammable gas generation
GT	Toxic gas generation
GF	Flammable gas generation
E	Explosion
P	Violent polymerization
S	Solubilization of toxic substances
U	May be hazardous but unknown

Source: Hatayama, et al., A Method for Determining the Compatibility of Hazardous Waste, U.S. EPA, 1980.

Appendix C
Chemical/Materials Compatibility Matrix

Sources:

Mellan, I., Corrosion Resistant Materials Handbook, Noyes Data Corporation, 1976.

Perry, R. and C. Chilton, Chemical Engineer's Handbook, McGraw-Hill, 1973.

Rabald, E., Corrosion Guide, Elsevier Scientific Publishing Company, 1968.

Shreir, L., Corrosion, Volumes 1 and 2, Newnes Publishing Company, 1976.

Staniar, W., Plant Engineering Handbook, McGraw-Hill, 1959.

Chemical Name	Chemical Class	Material: Mild Steel	Type 304 Stainless	Type 316 Stainless	Stainless 20 Alloys	Cast Iron	Silicon Cast Iron	Aluminum and Alloys	Nickel	Monel (65 Ni; 30 Cu)	Inconel (80 Ni; 14 Cr; 7 Fe)	Hastelloy B	Hastelloy C	Hastelloy D	Epoxy Resins	Furnace Resins	Glass	Phenolic Resins	PVC	Butyl Rubber	Neoprene	Ceramic	Saran	Polyethylene	Cement	Wood
Acetaldehyde	5	+	+	+	+	+	+	+	+	+	+	N	N	N	+	+	+	+	--	+	--	+	N	--	--	--
Acetamide	6	N	N	--	N	--	N	N	N	N	N	N	N	N	+	N	N	--	N	N	+	N	N	+	N	N
Acetic acid 10%	3	--	+	+	+	--	+	C	--	--	+	+	+	+	+	C	+	--	C	+	+	+	+	+	--	+
Acetic acid 100%	3	--	--	+	+	--	+	C	--	--	--	+	+	+	+	+	+	+	+	N	--	+	--	--	--	+
Acetic anhydride	38	--	+	+	+	+	+	C	+	+	+	+	+	+	+	--	+	+	--	--	--	+	--	--	--	--
Acetone	19	+	+	+	+	+	+	+	+	+	+	+	+	+	+	+	+	+	--	+	--	+	--	--	N	N
Acetophenone	19	+	+	+	+	+	N	+	+	+	+	N	N	N	N	N	+	N	--	+	--	+	N	--	N	N
Acrylonitrile	23, 34	N	N	N	N	N	N	N	N	N	N	N	N	N	--	N	N	--	N	--	--	N	N	N	N	N
Aldrin	17	N	+	+	N	N	N	N	N	N	N	N	N	N	+	+	N	N	N	N	N	N	N	N	N	N
Ally alcohol	4	N	N	N	N	N	N	+	N	N	N	N	N	N	N	N	N	+	C	N	N	N	N	--	N	N

L E G E N D

+ = Generally suitable
C = Conditionally suitable
-- = Generally unsuitable
N = Insufficient data

Chemical Name	Chemical Class	Material																								
		Mild Steel	Type 304 Stainless	Type 316 Stainless	Stainless 20 Alloys	Cast Iron	Silicon Cast Iron	Aluminum and Alloys	Nickel	Monel (65 Ni; 30 Cu)	Inconel (80 Ni; 14 Cr; 7 Fe)	Hastelloy B	Hastelloy C	Hastelloy D	Epoxy Resins	Furnace Resins	Glass	Phenolic Resins	PVC	Butyl Rubber	Neoprene	Ceramic	Saran	Polyethylene	Cement	Wood
Aluminum fluoride	21	C	N	N	N	C	N	+	+	N	N	N	N	N	+	+	N	+	+	+	+	+	+	+	+	+
Aluminum sulfate	21	--	--	C	+	--	+	--	C	C	C	+	+	+	+	+	+	+	+	+	+	+	+	+	--	+
Amino ethanol	4, 7	+	+	+	N	N	N	+	N	N	N	N	N	N	+	+	+	N	N	+	+	N	N	N	+	N
Ammonia, aq.	10	+	+	+	+	+	+	+	--	--	--	+	+	+	+	+	+	+	--	--	+	+	--	+	N	N
Ammonium fluoride	15	--	--	N	+	--	--	--	N	N	--	N	N	N	+	+	--	N	+	+	+	--	+	--	--	N
Ammonium hydroxide	10	+	+	+	+	N	N	--	--	--	+	N	N	N	+	+	C	+	+	+	+	+	--	+	N	N
Aniline	7	C	+	+	+	+	+	+	+	+	+	+	+	+	+	C	+	--	--	--	--	+	N	+	N	N
Beer		--	N	+	+	+	N	+	N	+	N	N	N	N	--	--	+	+	+	+	+	N	N	+	N	N
Benzene	16	+	+	+	+	+	+	+	+	+	+	+	+	+	--	C	+	+	--	--	--	+	--	--	--	N
Benzoic acid	3	--	+	+	+	--	+	+	+	+	+	+	+	+	+	+	+	--	+	+	+	+	+	+	--	N
Benzoly peroxide	27, 33	N	N	+	N	--	N	N	N	N	N	N	N	N	--	--	+	--	N	N	+	+	N	N	N	N
Benzyl chloride	17	--	N	+	--	--	+	--	+	+	N	+	+	N	--	C	+	C	--	--	--	+	N	N	--	--
Boric acid	1	--	+	+	+	--	+	+	C	+	+	+	+	+	+	+	+	+	+	+	+	+	+	+	--	N

LEGEND

+ = Generally suitable
C = Conditionally suitable
-- = Generally unsuitable
N = Insufficient data

Chemical Name	Chemical Class	Material																								
		Mild Steel	Type 304 Stainless	Type 316 Stainless	Stainless 20 Alloys	Cast Iron	Silicon Cast Iron	Aluminum and Alloys	Nickel	Monel (65 Ni; 30 Cu)	Inconel (80 Ni; 14 Cr; 7 Fe)	Hastelloy B	Hastelloy C	Hastelloy D	Epoxy Resins	Furnace Resins	Glass	Phenolic Resins	PVC	Butyl Rubber	Neoprene	Ceramic	Saran	Polyethylene	Cement	Wood
Butadiene	25, 34	+	+	+	N	N	N	+	N	N	N	N	N	N	N	N	N	N	N	N	--	N	N	--	N	N
Butane	26	+	+	+	N	N	N	+	N	N	N	N	N	N	N	N	N	--	N	--	--	N	N	+	N	N
Butyl acetate	13	+	+	+	+	+	+	+	+	+	+	+	+	+	C	--	+	--	--	--	--	+	--	--	--	N
Calcium hydroxide	10	+	+	+	+	+	--	C	+	+	+	+	+	+	+	+	C	--	+	+	+	+	+	+	N	N
Carbamide		N	N	+	N	N	N	+	N	N	N	N	N	N	+	+	+	N	N	+	+	N	N	N	N	N
Carbon disulfide	20	+	N	C	+	+	+	+	+	+	N	N	+	N	--	+	C	+	--	--	--	+	--	--	N	N
Carbon tetrachloride	17	--	--	C	+	+	+	C	+	+	+	+	+	+	+	+	+	+	+	--	--	+	+	--	N	N
Carbonic acid		N	+	+	+	N	+	C	--	N	+	+	+	N	+	+	+	+	+	+	+	+	+	+	--	N
Chloric acid	2, 35	--	--	--	+	N	N	--	--	--	N	N	N	N	N	N	+	N	N	N	N	N	N	+	N	N
Chloroacetone	17, 19	--	N	N	N	N	N	--	N	N	N	N	N	N	N	N	+	--	--	--	--	N	N	N	--	N

L E G E N D

+ = Generally suitable
C = Conditionally suitable
-- = Generally unsuitable
N = Insufficient data

Chemical Name	Chemical Class	Mild Steel	Type 304 Stainless	Type 316 Stainless	Stainless 20 Alloys	Cast Iron	Silicon Cast Iron	Aluminum and Alloys	Nickel	Monel (65 Ni; 30 Cu)	Inconel (80 Ni; 14 Cr; 7 Fe)	Hastelloy B	Hastelloy C	Hastelloy D	Epoxy Resins	Furnace Resins	Glass	Phenolic Resins	PVC	Butyl Rubber	Neoprene	Ceramic	Saran	Polyethylene	Cement	Wood
		Material																								
Chlorobenzene	17	+	+	+	+	+	+	+	+	+	+	N	N	N	C	+	+	--	--	--	--	+	--	--	--	N
Chlorocresol	17, 28	C	+	+	+	C	+	+	+	+	+	N	N	N	--	--	+	--	+	--	--	+	--	--	--	+
Chloroethanol	4, 17	+	+	+	N	N	N	+	N	N	N	N	N	N	+	+	+	N	--	+	--	N	N	N	N	N
Chloroform (dry)	17	+	+	+	+	+	+	+	+	+	+	+	+	+	+	C	+	+	--	--	--	+	--	--	N	N
Chlorosulfonic acid	1	--	N	C	+	--	+	C	+	+	+	+	+	+	--	--	+	--	+	--	--	N	N	N	N	N
Chromic acid 25%	2, 21, 35	--	--	--	C	--	+	--	--	--	N	N	+	N	--	--	+	--	+	+	--	+	+	C	--	N
Chromic acid 80%	2, 21, 35	+	--	--	+	--	+	--	--	--	N	N	+	N	--	--	+	--	+	--	--	+	+	C	--	N
Creosote	28	+	+	+	C	N	N	C	N	C	N	N	N	N	N	N	+	--	N	--	--	N	N	N	--	N
Cresol	28	+	+	+	+	+	+	+	+	+	N	N	N	N	N	+	+	--	--	--	--	N	N	--	--	N
Cumene	16	N	N	+	N	N	+	+	N	N	N	N	N	N	N	N	N	N	N	N	--	N	N	N	N	N
Cyclohexane	26	C	N	N	+	+	N	+	N	+	N	N	N	N	N	N	+	--	+	--	--	N	N	--	N	N
Cyclohexanone	19	--	N	N	N	N	N	N	N	N	N	N	N	N	N	N	+	N	--	--	--	N	N	--	N	N

LEGEND

+ = Generally suitable
C = Conditionally suitable
-- = Generally unsuitable
N = Insufficient data

Chemical Name	Chemical Class	Material: Mild Steel	Type 304 Stainless	Type 316 Stainless	Stainless 20 Alloys	Cast Iron	Silicon Cast Iron	Aluminum and Alloys	Nickel	Monel (65 Ni; 30 Cu)	Inconel (80 Ni; 14 Cr; 7 Fe)	Hastelloy B	Hastelloy C	Hastelloy D	Epoxy Resins	Furnace Resins	Glass	Phenolic Resins	PVC	Butyl Rubber	Neoprene	Ceramic	Saran	Polyethylene	Cement	Wood
Cyclohexanol	4	N	N	+	N	N	N	--	N	N	N	N	N	N	+	N	+	N	--	--	--	N	N	+	N	N
Diamine	8, 36	--	+	N	N	N	N	+	N	N	N	N	N	N	N	N	+	N	N	+	--	N	N	N	N	N
Dichloroacetone	17, 19	N	N	N	N	N	N	N	N	N	N	N	N	N	N	N	N	N	N	N	N	N	N	N	N	N
Dichloroethyl ether	14, 17	N	N	N	N	N	N	N	N	N	N	N	N	N	N	N	N	N	N	N	N	N	N	N	N	N
Dichloromethane	17	+	N	C	N	N	N	+	N	N	N	N	N	N	N	N	+	N	--	--	--	N	N	N	N	N
Diesel oil	32	+	+	+	+	+	+	+	N	N	N	N	N	N	N	N	N	N	N	--	N	N	N	N	N	N
Diethylamide	6	N	N	N	N	N	N	N	N	N	N	N	N	N	N	N	N	N	N	N	N	N	N	N	N	N
Diethylamine	7	C	+	+	+	N	+	C	+	+	+	N	N	N	N	N	+	N	--	+	+	+	N	N	N	N
Dimethylformamide	6	N	N	N	N	N	N	+	N	N	N	N	N	N	--	--	N	N	--	+	--	N	N	N	N	N
Dimethyl hydrazine	8	N	N	N	N	N	N	+	N	N	N	N	N	N	N	N	N	N	--	N	N	N	N	N	N	N
Dimethyl ketone	19	+	+	+	N	N	N	+	N	N	N	N	N	N	+	+	+	N	--	+	--	N	N	N	N	N
Dimethyl phthalate	13	N	N	N	N	N	N	+	N	N	N	N	N	N	N	N	N	N	N	+	--	N	N	--	N	N
Dioxane	14	+	+	N	N	N	N	C	N	N	N	N	N	N	N	N	+	N	--	--	--	N	N	N	--	--
Epichlorohydrin	17, 31	+	+	+	N	N	N	+	N	N	N	N	N	N	+	+	+	N	--	--	--	N	N	N	N	N
Ethanol (water free)	4	+	+	+	+	+	+	+	+	+	+	+	+	N	+	N	+	+	+	+	+	+	+	N	N	N

L E G E N D

+ = Generally suitable
C = Conditionally suitable
-- = Generally unsuitable
N = Insufficient data

		Material																									
Chemical Name	Chemical Class	Mild Steel	Type 304 Stainless	Type 316 Stainless	Stainless 20 Alloys	Cast Iron	Silicon Cast Iron	Aluminum and Alloys	Nickel	Monel (65 Ni; 30 Cu)	Inconel (80 Ni; 14 Cr; 7 Fe)	Hastelloy B	Hastelloy C	Hastelloy D	Epoxy Resins	Furnace Resins	Glass	Phenolic Resins	PVC	Butyl Rubber	Neoprene	Ceramic	Saran	Polyethylene	Cement	Wood
Ethyl acetate	13	+	+	+	+	+	+	+	+	+	+	+	+	+	+	+	+	+	--	C	--	+	--	--	--	+
Ethyl benzene	16	N	N	+	N	N	N	+	N	N	N	N	N	N	+	+	+	N	--	--	--	N	N	--	N	N
Ethylene chlorohydrin	4, 17	+	+	+	+	+	+	+	+	+	+	+	+	+	+	+	+	N	--	+	--	+	--	--	N	+
Ethylene diamine	7	N	N	+	N	N	N	C	N	N	N	N	N	N	N	N	+	N	N	+	+	N	N	N	N	N
Ethylene dichloride	17	+	+	+	+	N	+	C	+	+	C	+	N	N	+	C	+	+	--	--	--	+	--	--	N	N
Ethylene glycol	4	C	+	+	+	C	+	+	+	+	+	+	+	+	+	+	+	+	+	+	+	+	+	C	N	+
Ethylene glycol monobutyl ether	4, 14, 17	N	+	+	N	N	+	+	N	N	N	N	N	N	+	+	+	N	+	+	--	N	N	N	N	N
Ethyl ether	14	C	N	C	N	N	N	C	N	N	N	N	N	N	--	--	+	N	--	--	--	N	N	--	N	N
Ethyl mercaptan	20	--	N	+	N	N	N	+	N	N	N	N	N	N	--	--	+	N	N	--	--	N	N	N	N	N
Fatty acids		--	+	+	+	--	+	+	+	+	+	+	+	+	+	+	+	C	+	+	--	+	+	C	--	+
Fluosilicic acid	1, 15	--	--	--	+	--	--	--	N	C	N	N	N	N	+	+	--	N	C	+	+	N	N	C	N	N

LEGEND

+ = Generally suitable
C = Conditionally suitable
-- = Generally unsuitable
N = Insufficient data

Chemical Name	Chemical Class	Mild Steel	Type 304 Stainless	Type 316 Stainless	Stainless 20 Alloys	Cast Iron	Silicon Cast Iron	Aluminum and Alloys	Nickel	Monel (65 Ni; 30 Cu)	Inconel (80 Ni; 14 Cr; 7 Fe)	Hastelloy B	Hastelloy C	Hastelloy D	Epoxy Resins	Furnace Resins	Glass	Phenolic Resins	PVC	Butyl Rubber	Neoprene	Ceramic	Saran	Polyethylene	Cement	Wood
Formaldehyde	5	--	N	C	+	C	+	C	+	+	+	+	+	+	+	+	+	+	+	+	+	+	+	+	N	N
Formic acid	3	--	+	+	+	--	+	C	+	+	--	+	+	+	+	+	+	C	+	+	+	+	+	+	N	N
Freons	17	N	N	+	N	N	N	C	N	N	N	N	N	N	N	N	N	N	--	--	--	N	N	N	N	N
Furan	14	+	+	+	N	N	N	+	N	N	N	N	N	N	N	N	N	+	N	--	--	N	N	N	N	N
Furfural	5	+	+	+	+	+	+	+	+	+	N	N	N	N	--	C	+	--	--	+	--	N	N	--	N	N
Gasoline	32	+	+	+	+	+	+	+	+	+	+	+	+	+	+	+	+	+	C	--	--	+	+	--	N	N
Glycerine	4	+	+	+	+	+	+	+	+	+	+	C	+	+	N	N	+	+	+	+	C	+	+	+	--	N
Hydrazine	8, 36	--	+	+	+	--	N	+	--	--	--	--	--	--	+	+	--	+	+	C	--	N	N	N	N	--
Hydriodic acid	1	N	N	N	N	N	N	--	N	N	N	N	N	N	N	--	N	--	N	N	N	N	N	N	N	N
Hydrobromic acid	1, 38	--	--	--	--	--	+	--	N	--	N	+	N	N	+	C	+	C	+	+	--	+	+	+	--	N
Hydrochloric acid 15%	1	--	--	--	--	--	+	--	--	--	N	+	N	N	+	C	+	+	+	C	C	+	+	+	N	N
Hydrochloric acid 38%	1	--	--	--	--	--	+	--	--	--	N	+	N	N	+	C	+	--	+	--	--	+	+	+	N	N
Hydrocyanic acid (concentrated)	1, 11	C	+	+	+	C	+	+	+	+	+	+	+	+	+	+	+	C	+	+	--	+	+	+	N	N
Hydrofluoric acid 20%	1, 15	--	--	--	+	--	--	--	--	+	N	+	+	+	+	+	--	+	+	+	--	--	--	+	--	N
Hydrofluoric acid 75%	1, 15	--	--	--	+	--	--	--	--	+	N	+	+	+	--	--	--	--	C	+	--	--	--	+	--	N
Hydrofluorosilicic acid	15	--	N	C	+	N	N	--	N	N	N	N	N	N	+	+	--	N	+	+	+	N	N	N	N	N

LEGEND

+ = Generally suitable
C = Conditionally suitable
-- = Generally unsuitable
N = Insufficient data

Chemical Name	Chemical Class	Material																								
		Mild Steel	Type 304 Stainless	Type 316 Stainless	Stainless 20 Alloys	Cast Iron	Silicon Cast Iron	Aluminum and Alloys	Nickel	Monel (65 Ni; 30 Cu)	Inconel (80 Ni; 14 Cr; 7 Fe)	Hastelloy B	Hastelloy C	Hastelloy D	Epoxy Resins	Furnace Resins	Glass	Phenolic Resins	PVC	Butyl Rubber	Neoprene	Ceramic	Saran	Polyethylene	Cement	Wood
Hydrogen peroxide	27	--	C	C	C	--	C	+	C	C	+	N	+	N	N	--	C	C	+	+	+	+	N	+	C	--
Hydroquinone	28	C	+	C	N	--	N	C	+	+	+	N	N	N	+	+	+	+	+	C	C	+	N	+	N	N
Kerosene	32	+	+	+	+	+	+	+	+	+	+	+	+	+	+	+	+	+	+	--	--	+	+	--	N	N
Lactic acid	3	--	C	C	+	--	+	C	C	C	+	+	+	+	+	+	+	C	+	+	+	+	+	+	+	N
Malathion	29	--	N	+	N	N	N	N	N	N	N	N	N	N	N	N	N	N	N	N	N	N	N	N	N	N
Maleic acid	3	--	+	+	+	--	+	C	N	N	N	+	N	N	+	+	+	N	+	--	--	+	N	+	N	N
Methanol	4	+	+	+	+	+	+	+	+	+	+	+	+	N	+	+	+	+	+	+	+	+	+	+	+	+
Methyl acrylate	13, 34	+	+	+	N	N	N	+	N	N	N	N	N	N	N	N	+	N	C	C	C	N	N	N	N	N
Methyl amine	7	+	+	+	N	+	N	C	C	C	+	N	N	N	N	+	+	+	--	+	+	+	N	+	+	N
Methyl chloride	17	C	C	C	+	+	+	--	+	+	+	N	N	N	N	N	+	+	--	C	--	N	--	--	N	N
Methyl ethyl ketone	19	+	+	+	N	N	N	+	N	N	N	N	N	N	--	--	+	+	--	+	--	N	N	--	+	N
Methyl formate	13	+	+	+	N	N	N	+	N	N	N	N	N	N	N	N	N	N	C	+	+	N	N	N	N	N
Methyl isobutyl ketone	19	+	+	+	N	N	N	+	N	N	N	N	N	N	C	C	+	+	--	+	--	N	N	--	N	N

L E G E N D

+ = Generally suitable
C = Conditionally suitable
-- = Generally unsuitable
N = Insufficient data

Chemical Name	Chemical Class	Material: Mild Steel	Type 304 Stainless	Type 316 Stainless	Stainless 20 Alloys	Cast Iron	Silicon Cast Iron	Aluminum and Alloys	Nickel	Monel (65 Ni; 30 Cu)	Inconel (80 Ni; 14 Cr; 7 Fe)	Hastelloy B	Hastelloy C	Hastelloy D	Epoxy Resins	Furnace Resins	Glass	Phenolic Resins	PVC	Butyl.Rubber	Neoprene	Ceramic	Saran	Polyethylene	Cement	Wood
Monoethanolamine	4, 7	+	+	+	N	N	N	N	N	N	N	N	N	N	+	+	+	N	N	+	+	N	N	N	N	N
Naphtha (coal tar)	32	+	+	+	+	+	+	+	+	+	+	+	+	+	+	+	+	+	C	--	--	+	N	+	N	N
Naphthalene	16	N	N	N	+	+	N	+	N	N	N	N	N	N	N	N	N	+	--	--	--	N	+	--	N	N
Nitric acid 10%	2	--	+	+	+	--	+	--	--	--	--	--	+	--	--	N	+	C	+	+	--	+	+	+	--	--
Nitric acid 100%	2	--	C	C	+	--	+	C	--	--	--	--	C	--	--	--	+	--	--	--	--	+	--	--	--	--
Nitrobenzene	24	+	+	+	+	+	+	+	+	+	+	+	+	+	C	C	+	C	--	+	--	+	--	--	N	N
Nitrophenol	24, 28	C	+	+	N	C	N	+	+	+	+	N	N	N	+	+	+	+	--	N	N	N	N	N	C	+
Nitropropane	24	+	N	+	N	N	N	+	N	N	N	N	N	N	N	N	N	N	+	+	--	N	N	N	N	N
Nitrotoluene	24	+	+	+	+	+	+	+	+	+	+	N	N	N	N	N	+	N	N	--	--	+	N	N	+	+
Oleic acid	3	C	+	+	+	C	+	+	+	+	+	+	+	+	+	+	+	+	+	--	--	+	+	--	--	+
Oleum	2	C	C	C	+	C	N	+	--	--	N	N	+	N	--	--	+	N	--	--	+	N	N	--	--	--
Oxalic acid	3	--	C	C	+	--	+	C	+	+	+	+	+	+	+	+	+		+	+	+	+	+	+	+	N
Parathion	29	N	+	+	N	N	N	+	N	N	N	N	N	N	+	+	N	N	N	N	N	N	N	N	N	N
Pentachlorophenol	17, 28	+	+	+	N	N	N	N	N	N	N	N	N	N	N	N	N	N	N	N	N	N	N	N	N	N

LEGEND

+ = Generally suitable
C = Conditionally suitable
-- = Generally unsuitable
N = Insufficient data

Chemical Name	Chemical Class	Material																								
		Mild Steel	Type 304 Stainless	Type 316 Stainless	Stainless 20 Alloys	Cast Iron	Silicon Cast Iron	Aluminum and Alloys	Nickel	Monel (65 Ni; 30 Cu)	Inconel (80 Ni; 14 Cr; 7 Fe)	Hastelloy B	Hastelloy C	Hastelloy D	Epoxy Resins	Furnace Resins	Glass	Phenolic Resins	PVC	Butyl Rubber	Neoprene	Ceramic	Saran	Polyethylene	Cement	Wood
Perchloric acid	2	--	--	--	--	--	+	--	--	--	--	N	N	N	+	+	C	N	C	N	--	N	N	C	--	--
Phenol	28	+	+	+	+	+	+	+	+	+	+	+	+	+	+	+	+	--	--	+	--	+	--	--	+	+
Phosphoric acid 50%	1	--	+	+	+	--	+	--	C	+	C	N	+	N	+	+	+	C	+	+	+	--	+	+	--	N
Phosphoric acid 106%	1	--	C	+	+	--	+	--	C	--	N	N	+	N	+	+	+	--	+	+	N	--	N	N	--	N
Phthalic acid	3	C	+	+	+	C	+	+	+	+	+	+	+	N	+	N	+	N	N	+	N	+	+	N	N	N
Phthalic anhydride		C	+	+	+	C	+	+	+	+	+	+	+	N	+	N	+	N	N	+	N	+	+	N	N	+
Picric acid	24, 28, 33	--	+	+	+	--	+	--	--	--	--	+	+	+	N	N	+	--	--	N	--	+	--	+	N	N
Potassium cyanide	11	+	+	+	+	C	+	--	+	+	+	+	+	+	+	+	C	--	+	+	+	+	+	+	+	N
Propiolactone	13	N	+	+	N	N	N	N	N	N	N	N	N	N	N	N	N	N	N	N	N	N	N	N	N	N
Pyridine	7	+	+	+	+	+	+	+	N	N	N	N	N	N	N	N	+	+	--	--	--	+	N	+	+	+
Quinone	19	+	+	+	+	+	+	+	+	+	+	N	N	N	N	N	+	N	N	N	N	+	N	+	+	+
Resorcinol	28	N	N	N	N	N	N	+	N	N	N	N	N	--	N	N	N	N	N	N	N	N	N	N	N	N

LEGEND

+ = Generally suitable
C = Conditionally suitable
-- = Generally unsuitable
N = Insufficient data

Chemical Name	Chemical Class	Material																								
		Mild Steel	Type 304 Stainless	Type 316 Stainless	Stainless 20 Alloys	Cast Iron	Silicon Cast Iron	Aluminum and Alloys	Nickel	Monel (65 Ni; 30 Cu)	Inconel (80 Ni; 14 Cr; 7 Fe)	Hastelloy B	Hastelloy C	Hastelloy D	Epoxy Resins	Furnace Resins	Glass	Phenolic Resins	PVC	Butyl Rubber	Neoprene	Ceramic	Saran	Polyethylene	Cement	Wood
Salycilic acid	3	C	+	+	+	C	+	C	+	+	+	N	N	N	+	+	+	+	--	+	C	+	+	+	N	+
Silver nitrate	21, 35	--	+	+	+	+	+	--	--	--	+	+	+	+	+	+	+	+	+	+	+	+	+	+	N	N
Soap solutions		C	C	C	+	C	+	C	+	C	+	+	+	+	+	+	+	+	+	+	+	+	N	+	N	N
Sodium carbonate	10	+	+	+	+	+	+	--	+	+	+	+	+	+	+	+	+	+	+	+	+	+	+	+	C	N
Sodium chloride		+	C	C	+	C	+	C	+	+	C	+	+	+	+	+	+	+	+	+	+	+	+	+	+	+
Sodium cyanide	11	+	+	+	+	+	+	--	--	--	--	N	N	N	+	+	C	+	+	+	+	+	--	+	+	+
Sodium hydroxide	10	+	+	+	+	+	--	--	+	+	+	+	N	+	+	+	C	--	+	+	+	--	--	+	+	N
Sodium hypochlorite	10, 35	--	--	--	C	--	+	--	C	C	C	--	+	--	C	--	+	--	+	+	--	+	+	+	N	N
Styrene	16, 25, 34	N	N	C	N	N	N	+	N	N	N	N	N	N	N	N	+	--	--	--	--	N	N	--	N	N
Sulfuric acid (0 to 30%)	2, 38	--	--	C	+	--	C	--	--	+	--	+	+	+	+	+	+	--	+	+	+	+	+	+	N	N
Sulfuric acid (50%)	2, 38	--	--	--	+	--	+	--	--	+	--	+	+	+	+	+	+	--	+	+	+	+	--	+	N	N
Sulfuric acid (95%)	2, 38	C	--	C	+	+	+	--	--	C	--	+	+	+	--	--	+	--	--	--	--	+	--	--	N	N
Sulfur trioxide (dry)	35, 38	+	+	+	+	+	--	+	--	--	--	N	N	N	+	N	+	N	+	+	--	+	+	+	N	N
Tetrachloroethane	17	N	N	+	N	N	N	--	N	N	N	N	N	N	+	+	+	N	--	--	--	N	N	N	N	N
Tetraethyl lead	21	N	N	+	N	N	N	N	N	N	N	N	N	N	N	N	N	N	+	--	N	N	N	+	N	N

LEGEND

+ = Generally suitable
C = Conditionally suitable
-- = Generally unsuitable
N = Insufficient data

Chemical Name	Chemical Class	Mild Steel	Type 304 Stainless	Type 316 Stainless	•Stainless 20 Alloys	Cast Iron	Silicon Cast Iron	Aluminum and Alloys	Nickel	Monel (65 Ni; 30 Cu)	Inconel (80 Ni; 14 Cr; 7 Fe)	Hastelloy B	Hastelloy C	Hastelloy D	Epoxy Resins	Furnace Resins	Glass	Phenolic Resins	PVC	Butyl Rubber	Neoprene	Ceramic	Saran	Polyethylene	Cement	Wood
		Material																								
Tetrahydrofuran	14	N	N	+	+	N	N	N	N	N	N	N	+	N	--	N	N	N	--	N	--	+	N	--	N	N
Toluene	16	+	+	+	+	+	+	+	+	+	+	+	+	+	+	+	+	+	--	--	--	+	+	--	C	C
Transformer oil		N	N	+	N	N	N	+	N	N	N	N	N	N	+	+	+	+	--	--	+	N	N	N	N	N
Trichloroethylene	17	--	--	+	+	+	+	+	+	+	+	+	+	+	+	N	+	+	--	C	--	+	--	--	N	N
Turpentine	32	N	N	+	N	N	N	+	N	N	N	N	N	N	+	+	+	N	N	--	--	N	N	+	N	N
Urea		N	N	+	N	N	N	+	N	N	N	N	N	N	+	+	+	N	+	+	+	N	N	+	N	N
Xylene	16	+	+	N	N	N	N	+	N	N	N	N	N	N	--	--	+	+	N	--	--	N	N	--	--	N
Zinc chloride	21	--	C	C	+	--	+	C	+	+	+	+	+	N	+	+	+	+	+	+	+	+	+	+	N	C

LEGEND

+ = Generally suitable
C = Conditionally suitable
-- = Generally unsuitable
N = Insufficient data

Appendix D
Hazardous Substance/Countermeasure Matrix

Appendix D consists of matrix of countermeasures recommended for treating hazardous substance spills. Chemicals are listed in alphabetical order in the first column. The second column identifies each compound's EPA Toxicity Classification, based on LC_{50} toxic concentrations, as follows:

Category	Toxicity Range
A	$LC_{50} \leq 1$ ppm
B	1 ppm $< LC_{50} \leq 10$ ppm
C	10 ppm $< LC_{50} \leq 100$ ppm
D	100 ppm $< LC_{50} \leq 500$ ppm

The third and fourth columns list, respectively, the density and physical form (solid or liquid) of the pure hazardous substance. The physical/chemical properties of a chemical discharge (solubility, density, volatility, and ability to disperse in water) must be considered in estimating its potential to harm the environment. Column five identifies the P/C/D category, which takes into account physical/chemical properties. The P/C/D categories are as follows:

IVF - Insoluble Volatile Floater
INF - Insoluble Non-volatile Floater
IS - Insoluble Sinker
SM - Soluble Mixer
P - Precipitator
SF - Soluble Floater
M - Miscible
SS - Soluble Sinker

The remaining columns of the matrix indicate which categories of countermeasures are effective for controlling hazardous substances discharged on the ground or into water.

MATERIAL	EPA CATE-GORY	DENSITY	PHYSICAL FORM	P/C/D CATE-GORY	MASS TRANSFER MEDIA			NEUTRALIZING AGENT		PRECIPI-TATING AGENT	BIOLOGICAL TREATMENT AGENT	GELLING AGENT	ABSORBING AGENT	OXIDIZING AGENT	DISPERSING AGENT
					ACTIVA-TED CARBON	CATIONIC RESIN	ANIONIC RESIN	ACID	BASE						
ACETALDEHYDE	C	0.783	L	M	●				●			●	●		
ACETIC ACID	C	1.049	L	M	●				●		●	●	●		
ACETIC ANHYDRIDE	C	1.083	L	SF	●				●		●	●	●		●
ACETONE CYANOHYDRIN	C	0.90	L	SF	●							●	●		●
ACETYL BROMIDE	D	1.52	L	SS	●		●					●	●		
ACETYL CHLORIDE	D	1.11	L	SS	●				●			●	●		
ACROLEIN	A	0.839	L	SF	●						●	●	●		●
ACRYLONITRILE	C	0.807	L	SF	●							●	●		●
ADIPONITRILE	D	0.95	L	SF	●							●	●		●
ALDRIN	A	1.65	S	IS	●										
ALLYL ALCOHOL	B	0.854	L	M	●						●	●	●		
ALLYL CHLORIDE	C	0.9	L	IVF	●						●	●	●		●
ALUMINUM FLUORIDE	D	2.88	S	P	●	●	●			●					
ALUMINUM SULFATE	D	1.69	S	P	●					●					
AMMONIA	C	0.60	L	SF	●	●		●				●			●
AMMONIUM ACETATE	D	1.073	S	SM	●	●									
AMMONIUM BENZOATE	D	1.26	S	SS	●	●	●								
AMMONIUM BICARBONATE	D	1.58	S	SS	●	●									
AMMONIUM BICHROMATE	D	2.15	S	SS	●	●	●			●					
AMMONIUM BIFLUORIDE	D	1.21	S	SS	●	●	●			●					
AMMONIUM BISULFITE	D	–	S	SS	●	●								●	
AMMONIUM BROMIDE	D	2.43	S	SS	●	●	●								
AMMONIUM CARBAMATE	D	–	S	SS	●	●	●								
AMMONIUM CARBONATE	S	–	S	SM	●	●									
AMMONIUM CHLORIDE	D	1.53	S	SS	●	●									
AMMONIUM CHROMATE	D	1.91	S	SS	●	●	●								
AMMONI'JM CITRATE	D	–	S	SS	●	●									
AMMONIUM FLUOBORATE	D	1.85	S	SS	●	●	●								
AMMONIUM FLUORIDE	D	1.31	S	SM	●	●	●			●					
AMMONIUM HYDROXIDE	C	0.9	S/L	M	●	●		●				●			
AMMONIUM HYPOPHOSPHITE	D	–	S	SS	●	●									
AMMONIUM IODIDE	D	2.56	S	SM	●	●	●								
AMMONIUM NITRATE	D	1.66	S	SM	●	●	●								
AMMONIUM OXALATE	D	1.50	S	SS	●	●	●								
AMMONIUM PENTABORATE	D	–	S	SS	●	●	●								
AMMONIUM PERSULFATE	D	1.98	S	SS	●	●									
AMMONIUM SILICOFLUORIDE	C	2.01	S	SS	●	●	●								

MATERIAL	EPA CATEGORY	DENSITY	PHYSICAL FORM	P/C/D CATEGORY	MASS TRANSFER MEDIA: ACTIVATED CARBON	MASS TRANSFER MEDIA: CATIONIC RESIN	MASS TRANSFER MEDIA: ANIONIC RESIN	NEUTRALIZING AGENT: ACID	NEUTRALIZING AGENT: BASE	PRECIPITATING AGENT	BIOLOGICAL TREATMENT AGENT	GELLING AGENT	ABSORBING AGENT	OXIDIZING AGENT	DISPERSING AGENT
AMMONIUM SULFAMATE	D	–	S	SM	●	●	●								
AMMONIUM SULFIDE	D	1.02	S	SS	●	●	●							●	
AMMONIUM SULFITE	D	1.41	S	SS	●	●	●							●	
AMMONIUM TARTRATE	D	1.61	S	SS	●	●	●								
AMMONIUM THIOCYANATE	D	1.31	S	SM	●	●	●								
AMMONIUM THIOSULFATE	D	–	S	SM	●	●	●								
AMYL ACETATE	C	0.88	L	INF	●						●	●	●		●
ANILINE	C	1.022	L	SS	●						●	●	●		●
ANTIMONY PENTACHLORIDE	C	2.34	S	P	●	●				●					
ANTIMONY PENTAFLUORIDE	C	2.99	S	P	●	●	●			●					
ANTIMONY POTASSIUM TARTRATE	C	2.6	S	P	●	●	●			●					
ANTIMONY TRIBROMIDE	C	4.14	S	P	●	●	●			●					
ANTIMONY TRICHLORIDE	C	3.14	S	P	●	●				●					
ANTIMONY TRIFLUORIDE	C	4.38	S	P	●	●	●			●					
ANTIMONY TRIOXIDE	C	5.2	S	P	●	●				●					
ARSENIC ACID	C	2 - 2.5	S	P	●		●		●	●		●	●		
ARSENIC DISULFIDE	C	3.4	S	IS	●	●	●			●		●	●		
ARSENIC PENTOXIDE	B	4.09	S	P	●	●				●					
ARSENIC TRICHLORIDE	C	2.16	S	P	●	●				●					
ARSENIC TRIOXIDE	B	3.89	S	P	●	●	●			●					
ARSENIC TRISULFIDE	B	3.43	S	IS	●	●	●			●					
BARIUM CYANIDE	A	–	S	SS	●	●	●			●				●	
BENZENE	C	0.879	L	INF	●							●	●		●
BENZOIC ACID	D	1.266	S	SS	●				●			●	●		
BENZONTRILE	C	1.01	L	SS	●							●	●		
BENZOYL CHLORIDE	D	1.20	L	SS	●							●	●		
BENZYL CHLORIDE	D	1.09	L	IS	●							●	●		
BERYLLIUM CHLORIDE	D	1.90	S	P	●	●				●					
BERYLLIUM FLUORIDE	C	1.99	S	P	●	●	●			●					
BERYLLIUM NITRATE	C	1.56	S	P	●	●				●					
BUTYL ACETATE	C	0.89	L	SF	●						●	●	●		●
BUTYLAMINE	C	0.74	L	M	●							●	●		
BUTYRIC ACID	D	1.00	L	M	●		●		●		●	●	●		●
CADMIUM ACETATE	A	2.01	S	SS	●	●				●					
CADMIUM BROMIDE	A	5.19	S	P	●	●	●			●					
CADMIUM CHLORIDE	A	4.05	S	P	●	●				●					
CALCIUM ARSENATE	C	3.0	S	IS	●		●			●					

MATERIAL	EPA CATE-GORY	DENSITY	PHYSICAL FORM	P/C/D CATE-GORY	MASS TRANSFER MEDIA: ACTIVA-TED CARBON	MASS TRANSFER MEDIA: CATIONIC RESIN	MASS TRANSFER MEDIA: ANIONIC RESIN	NEUTRALIZING AGENT: ACID	NEUTRALIZING AGENT: BASE	PRECIPI-TATING AGENT	BIOLOGICAL TREATMENT AGENT	GELLING AGENT	ABSORBING AGENT	OXIDIZING AGENT	DISPERSING AGENT
CALCIUM ARSENITE	C	–	S	SS	●		●								
CALCIUM CARBIDE	D	2.2	S	P	●										
CALCIUM CHROMATE	D	2.89	S	SS	●		●								
CALCIUM CYANIDE	A	–	S	SS	●		●							●	
CALCIUM DODECYLBENZENE-SULFONATE	B	–	S	SS	●		●								
CALCIUM HYDROXIDE	D	2.504	S	SS				●							
CALCIUM HYPOCHLORITE	A	2.35	S	SM	●				●						
CALCIUM OXIDE	D	3.40	S	SM	●										
CAPTAN	A	1.5	S	SS	●								●		
CARBARYL	B	–	S	SS	●								●		
CARBON DISULFIDE	C	1.26	L	SS	●							●	●		
CHLORDANE	A	1.59	L	IS	●							●	●		
CHLORINE	A	3.2	L	SF	●								●		
CHLOROBENZENE	B	1.1	L	IS	●							●	●		
CHLOROFORM	B	1.5	L/G	IS	●							●	●		
CHLOROSULFONIC ACID	C	1.8	L	SS	●		●		●			●	●		
CHROMIC ACETATE	D	–	S	SS	●	●				●					
CHROMIC ACID	D	2.7	L	SM	●	●			●	●		●	●		
CHROMIC SULFATE	D	1.7	S	SS	●	●				●					
CHROMOUS CHLORIDE	D	2.87	S	IS	●	●				●					
CHROMYL CHLORIDE	D	1.91	S	SS	●	●				●					
COBALTOUS BROMIDE	C	2.47	S	P	●	●	●			●					
COBALTOUS FLUORIDE	C	4.46	S	P	●	●	●			●					
COBALTOUS FORMATE	C	2.13	S	P	●	●				●					
COBALTOUS SULFAMATE	C	–	S	P	●	●				●					
COUMAPHOS	A	–	S	SS	●								●		
CRESOL	B	1.0	S	SS	●						●	●	●		●
CUPRIC ACETATE	B	1.9	S	P	●	●				●					
CUPRIC ACETOARSENITE	B	–	S	IS	●	●	●			●					
CUPRIC CHLORIDE	B	3.39	S	P	●	●				●					
CUPRIC FORMATE	B	1.83	S	P	●	●				●					
CUPRIC GLYCINATE	B	–	S	P	●	●				●					
CUPRIC LACTATE	B	–	S	P	●	●				●					
CUPRIC NITRATE	B	2.32	S	P	●	●				●					
CUPRIC OXALATE	B	–	S	IS	●	●	●			●					
CUPRIC SUBACETATE	B	1.9	S	P	●	●				●					

MATERIAL	EPA CATE-GORY	DENSITY	PHYSICAL FORM	P/C/D CATE-GORY	MASS TRANSFER MEDIA			NEUTRALIZING AGENT		PRECIPI-TATING AGENT	BIOLOGICAL TREATMENT AGENT	GELLING AGENT	ABSORBING AGENT	OXIDIZING AGENT	DISPERSING AGENT
					ACTIVA-TED CARBON	CATIONIC RESIN	ANIONIC RESIN	ACID	BASE						
CUPRIC SULFATE	B	2.28	S	P	•	•				•					
CUPRIC SULFATE AMMONIATED	B	–	S	P	•	•				•					
CUPRIC TARTRATE	B	–	S	IS	•	•				•					
CUPROUS BROMIDE	B	4.72	S	IS	•	•	•			•					
CYANOGEN CHLORIDE	A	1.186	G	SS	•	•									
CYCLOHEXANE	C	0.779	L	INF	•						•	•	•		•
2, 4-D ACID	B	0.82	–	IS									•		
2, 4-D ESTERS	B	–	–	IS									•		
DALAPON	B	1.38	L	SS	•							•	•		
DDT	A	–	S	IS	•							•	•		
DIAZINON	A	1.116	L	IS	•							•	•		
DICAMBA	C	–	S	SS	•								•		
DICHLOBENIL	C	–	S	SS	•								•		
DICHLONE	A	–	S	SS	•							•	•		
DICHLORVOS	A	–	L	SS	•								•		
DIELDRIN	A	1.75	S	SS	•								•		
DIETHYLAMINE	C	0.71	L	SF	•						•	•	•		•
DIMETHYLAMINE	C	0.68	L	SF	•						•	•	•		•
DINITROBENZENE	C	1.54	L	SS	•						•	•	•		•
DINITROPHENOL	B	1.68	L	SS	•						•	•	•		•
DIQUAT	C	–	S	SS	•								•		
DISULFOTON	A	1.14	L	SS	•								•		
DIURON	B	–	S	SS								•	•		
DODECYLBENZENESULFONIC ACID	B	–	L	SS	•		•				•	•	•		•
DURSBAN	B	–	–	SS	•								•		
ENDOSULFAN	A	–	S	SS	•								•		
ENDRIN	A	–	S	IS	•								•		
ETHION	A	1.22	L	SS	•							•	•		
ETHYLBENZENE	C	0.958	L	INF	•						•	•	•		•
ETHYLENEDIAMINE	C	0.96	L	SF	•						•	•	•		•
EDTA	D	–	S	IS	•						•		•		•
FERRIC AMMONIUM CITRATE	C	–	S	P	•	•	•			•					
FERRIC AMMONIUM OXALATE	C	–	S	P	•	•				•					
FERRIC CHLORIDE	C	2.89	S	P	•	•				•					
FERRIC FLUORIDE	C	3.52	S	P	•	•	•			•					
FERRIC NITRATE	C	1.68	S	P	•	•				•					
FERRIC SULFATE	C	2.0	S	P	•	•				•					

MATERIAL	EPA CATEGORY	DENSITY	PHYSICAL FORM	P/C/D CATEGORY	MASS TRANSFER MEDIA: ACTIVATED CARBON	MASS TRANSFER MEDIA: CATIONIC RESIN	MASS TRANSFER MEDIA: ANIONIC RESIN	NEUTRALIZING AGENT: ACID	NEUTRALIZING AGENT: BASE	PRECIPITATING AGENT	BIOLOGICAL TREATMENT AGENT	GELLING AGENT	ABSORBING AGENT	OXIDIZING AGENT	DISPERSING AGENT
FERROUS AMMONIUM SULFATE	C	1.87	S	P	•	•	•			•					
FERROUS CHLORIDE	C	1.93	S	P	•	•				•					
FERROUS SULFATE	C	1.899	S	P	•	•				•					
FORMALDEHYDE	C	0.815	L	M	•		•				•	•	•		•
FORMIC ACID	C	1.22	L	M	•				•		•	•	•		•
FUMARIC ACID	D	1.635	L	SS	•				•		•	•	•		•
FURFURAL	C	1.15	L	SS	•						•	•	•		•
GUTHION	A	1.44	L	IS	•							•	•		
HEPTACHLOR	A	1.58	S	IS	•								•		
HYDROCHLORIC ACID	D	1.00	L	SS	•		•		•			•	•		
HYDROFLUORIC ACID	D	1.15	L	M	•		•		•			•	•		
HYDROGEN CYANIDE	A	0.70	L/G	M	•		•		•			•	•	•	
HYDROXYLAMINE	D	1.23	S	SS	•								•		
ISOPRENE	C	0.681	L	IVF	•						•	•	•		•
ISOPROPANOLAMINE DODECYL-BENZENESULFONATE	B	0.90	L	SS	•						•	•	•		•
KELTHANE	C	–	–	IS	•								•		
LEAD ACETATE	D	2.25	S	P	•	•	•			•					
LEAD ARSENATE	D	7.8	S	IS	•	•	•			•					
LEAD CHLORIDE	D	5.85	S	P	•	•				•					
LEAD FLUBORATE	D	–	S	P	•	•	•			•					
LEAD FLUORIDE	C	8.2	S	IS	•	•	•			•					
LEAD IODIDE	D	6.16	S	IS	•	•	•			•					
LEAD NITRATE	D	4.53	S	P	•	•	•			•					
LEAD STERATE	D	1.4	S	P	•	•				•					
LEAD SULFATE	D	6.2	S	I$	•	•				•					
LEAD SULFIDE	C	7.1	S	IS	•	•	•			•				•	
LEAD TETRAACETATE	D	2.23	S	P	•	•				•					
LEAD THIOCYANATE	D	3.8	S	IS	•	•				•					
LEAD THIOSULFATE	D	5.18	S	IS	•	•				•					
LEAD TUNGSTATE	D	8.24	S	IS	•	•	•			•					
LINDANE	A	1.87	S	SS	•								•		
LITHIUM BICHROMATE	D	2.34	S	SM	•	•	•			•					
LITHIUM CHROMATE	D	–	S	SM	•	•	•			•					
MALATHION	A	1.23	L	SS								•	•		
MALEIC ACID	D	1.59	S	SS	•				•		•		•		•
MALEIC ANHYDRIDE	D	0.934	S	SF	•				•		•		•		•

MATERIAL	EPA CATE-GORY	DENSITY	PHYSICAL FORM	P/C/D CATE-GORY	MASS TRANSFER MEDIA: ACTIVA-TED CARBON	MASS TRANSFER MEDIA: CATIONIC RESIN	MASS TRANSFER MEDIA: ANIONIC RESIN	NEUTRALIZING AGENT: ACID	NEUTRALIZING AGENT: BASE	PRECIPI-TATING AGENT	BIOLOGICAL TREATMENT AGENT	GELLING AGENT	ABSORBING AGENT	OXIDIZING AGENT	DISPERSING AGENT
MERCURIC ACETATE	A	3.25	S	P	●	●				●			●		
MERCURIC CYANIDE	A	4.09	S	P	●	●	●			●			●	●	
MERCURIC NITRATE	A	4.3	S	P	●	●				●			●		
MERCURIC SULFATE	A	6.47	S	P	●	●				●			●		
MERCURIC THIOCYANATE	A	–	S	IS	●	●	●			●			●		
MERCUROUS NITRATE	A	4.79	S	P	●	●				●			●		
METHOXYCHLOR	A	1.41	S	IS	●								●		
METHYL MERCAPTAN	B	0.87	L/G	INF	●							●	●		●
METHYL METHACRYLATE	D	0.936	L	INF	●							●	●		●
METHYL PARATION	B	1.358	L	IS	●							●	●		
MEVINPHOS	A	–	L	M	●							●	●		
MONOETHYLAMINE	C	1.01	–	M	●								●		●
MONOMETHYLAMINE	C	–	–	SF	●								●		●
NALED	A	–	S/L	IS	●							●	●		
NAPTHALENE	B	1.162	S	IS	●								●		
NAPTHENIC ACID	A	1.4	S	SS	●							●	●		
NICKEL AMMONIUM SULFATE	D	1.92	S	P	●	●				●					
NICKEL CHLORIDE	D	3.55	S	P	●	●				●					
NICKEL FORMATE	C	2.15	S	P	●	●				●					
NICKEL HYDROXIDE	C	4.36	S	IS	●	●				●					
NICKEL NITRATE	D	2.05	S	P	●	●				●					
NICKEL SUFLATE	D	1.948	S	P	●	●				●					
NITRIC ACID	C	1.502	L	M	●				●			●			
NITROBENZENE	D	1.19	L	SS	●							●	●		
NITROGEN DIOXIDE	C	1.448	L/G	M	●										
NITROPHENOL	B	1.4	L	SS	●						●	●	●		●
PARAFORMALDEHYDE	C	1.46	S	SS	●						●		●		●
PARATHION	A	1.26	L	IS	●							●	●		
PENTACHLOROPHENOL	A	1.978	S	IS	●							●	●		
PHENOL	B	1.071	S	SS	●		●				●	●	●		
PHOSGENE	D	1.392	G/L	SS	●								●		●
PHOSPHORIC ACID	D	1.834	L	M	●				●			●	●		
PHOSPHOROUS	A	1.8 - 2.7	S	IS											
PHOSPHOROUS OXYCHLORIDE	D	1.67	L	SS	●	●						●			
PHOSPHOROUS PENTASULFIDE	C	2.03	S	SS	●										
PHOSPHOROUS TRICHLORIDE	D	1.574	S	SS	●	●	●								
POLYCHLORINATED BIPHENYLS	A	–	S	IS									●		

MATERIAL	EPA CATE-GORY	DENSITY	PHYSICAL FORM	P/C/D CATE-GORY	MASS TRANSFER MEDIA			NEUTRALIZING AGENT		PRECIPI-TATING AGENT	BIOLOGICAL TREATMENT AGENT	GELLING AGENT	ABSORBING AGENT	OXIDIZING AGENT	DISPERSING AGENT
					ACTIVA-TED CARBON	CATIONIC RESIN	ANIONIC RESIN	ACID	BASE						
POTASSIUM ARSENATE	C	2.87	S	P	•		•								
POTASSIUM ARSENITE	C	–	S	P	•		•								
POTASSIUM BICHROMATE	D	2.68	S	SS	•		•								
POTASSIUM CHROMATE	D	2.73	S	SS	•		•								
POTASSIUM CYANIDE	A	1.52	S	SS	•		•							•	
POTASSIUM HYDROXIDE	C	2.04	S	SM	•			•							
POTASSIUM PERMANGANATE	B	2.7	S	SS	•		•								
PROPRIONIC ACID	D	0.993	L	M	•				•		•	•	•		•
PROPRIONIC ANHYDRIDE	D	1.013	L	M	•				•		•	•	•		•
PROPYL ALCOHOL	D	0.8	L	M	•						•	•	•		•
PYRETHRINS	C	–	L	SS								•	•		
QUINOLINE	A	1.09	L	SS	•							•	•		•
RESORCINOL	B	1.27	S	SS	•						•		•		
SELENIUM OXIDE	C	3.954	S	SS	•	•				•					
SODIUM	C	0.971	S	SS											
SODIUM ARSENATE	C	1.76	S	SS	•		•								
SODIUM ARSENITE	C	1.87	S	SS	•		•								
SODIUM BICHROMATE	D	2.52	S	SM	•	•									
SODIUM BIFLUORIDE	D	2.08	S	SS	•		•			•					
SODIUM BISULFITE	D	1.48	S	SS	•		•							•	
SODIUM CHROMATE	D	1.483	S	SS	•		•								
SODIUM CYANIDE	A	1.48	S	SS	•		•							•	
SODIUM DODECYLBENZENE-SULFONATE	B	–	S	SS	•		•				•		•		•
SODIUM FLUORIDE	D	2.78	S	SS	•		•			•					
SODIUM HYDROSULFIDE	D		S	SS	•		•						•		
SODIUM HYDROXIDE	C	2.13	L	SS	•			•				•			
SODIUM HYPOCHLORITE	A	–	S	SM	•		•								
SODIUM METHYLATE	C	2.4	S	SS	•		•				•		•		•
SODIUM NITRITE	B	2.17	S	SS	•										
SODIUM PHOSPHATE MONOBASIC	D	2.04	S	SS	•										
SODIUM PHOSPHATE DIBASIC	D	2.06	S	SM	•										
SODIUM PHOSPHATE TRIBASIC	D	1.5	S	SS											
SODIUM SELENITE	C	1.63	S	SS			•								
SODIUM SULFIDE	C	1.856	S	SS	•		•			•				•	
STANNOUS FLUORIDE	D	2.79	S	SS		•	•			•					
STRONTIUM CHROMATE	D	–	S	IS	•	•	•			•					

MATERIAL	EPA CATEGORY	DENSITY	PHYSICAL FORM	P/C/D CATEGORY	MASS TRANSFER MEDIA			NEUTRALIZING AGENT		PRECIPITATING AGENT	BIOLOGICAL TREATMENT AGENT	GELLING AGENT	ABSORBING AGENT	OXIDIZING AGENT	DISPERSING AGENT
					ACTIVATED CARBON	CATIONIC RESIN	ANIONIC RESIN	ACID	BASE						
STRYCHNINE	C	1.36	S	SS	●								●		
STYRENE	C	0.909	L	INF	●							●	●		●
SULFURIC ACID	C	1.834	L	M	●				●			●	●		
SULFUR MONOCHLORIDE	D	1.69	S	SS	●				●						
2,4,5-T ACID	A	–	S	IS									●		
2,4,5-T ESTERS	A	–	S	IS									●		
TDE	A	–	S	IS	●								●		
TETRAETHYL LEAD	A	1.659	L	IS	●							●	●		
TETRAETHYL PYROPHOSPHATE	B	1.2	L	M	●							●			
TOLUENE	C	0.86	L	INF	●						●	●	●		●
TOXAPHENE	A	1.66	L	IS	●							●	●		
TRICHLORFON	B	1.73	S	SS	●								●		
TRICHLOROPHENOL	A	1.1	L	IS	●							●	●		
TRIETHANOLAMINE DODECYL-BENZENESULFONATE	B	–	L	SS	●							●	●		
TRIETHYLAMINE	C	1.13	L	SF	●						●	●	●		●
TRIMETHYLAMINE	C	0.66	L	SF	●						●	●	●		●
URANIUM PEROXIDE	D	2.5	S	IS	●	●				●					
URANYL ACETATE	D	2.89	S	P	●	●				●					
URANYL NITRATE	D	2.80	S	P	●	●				●					
URANYL SULFATE	D	3.28	S	P	●	●				●					
VANADIUM PENTOXIDE	C	3.36	S	P	●	●				●					
VANADYL SUFATE	C	–	S	P	●	●				●					
VINYL ACETATE	C	0.94	S	SF	●						●	●	●		●
XYLENE	C	0.86	L	INF	●							●	●		●
XYLENOL	C	1.02	L	SS	●							●	●		
ZECTRAN	C	–	–	SS	●								●		
ZINC ACETATE	C	1.735	S	P	●	●				●					
ZINC AMMONIUM CHLORIDE	C	1.80	S	P	●	●	●			●					
ZINC BICHROMATE	C	–	S	P	●	●	●			●					
ZINC BORATE	C	3.64	S	P	●	●	●			●					
ZINC BROMIDE	C	4.22	S	P	●	●	●			●					
ZINC CARBONATE	C	4.42	S	IS	●	●				●					
ZINC CHLORIDE	C	2.907	S	P	●	●				●					
ZINC CYANIDE	A	1.85	S	IS	●	●	●			●				●	
ZINC FLURODIE	C	4.84	S	P	●	●	●			●					
ZINC FORMATE	C	2.21	S	P	●	●				●					

MATERIAL	EPA CATEGORY	DENSITY	PHYSICAL FORM	P/C/D CATEGORY	MASS TRANSFER MEDIA			NEUTRALIZING AGENT		PRECIPITATING AGENT	BIOLOGICAL TREATMENT AGENT	GELLING AGENT	ABSORBING AGENT	OXIDIZING AGENT	DISPERSING AGENT
					ACTIVATED CARBON	CATIONIC RESIN	ANIONIC RESIN	ACID	BASE						
ZINC HYDROSULFITE	C	–	S	P	●	●	●			●				●	
ZINC NITRATE	C	2.07	S	P	●	●				●					
ZINC PHENOLSULFONATE	C	–	S	P	●	●	●			●					
ZINC PHOSPHIDE	C	4.55	S	IS	●	●	●			●					
ZINC POTASSIUM CHROMATE	C	–	S	IS	●	●	●			●					
ZINC SULICOFLUORIDE	C	2.1	S	P	●	●	●			●					
ZINC SULFATE	C	3.54	S	P	●	●				●					
ZINC SULFATE MONOHYDRATE	C	3.28	S	P	●	●				●					
ZIRCONIUM ACETATE	D	–	S	P	●	●				●					
ZIRCONIUM NITRATE	D	–	S	P	●	●				●					
ZIRCONIUM OXYCHLORIDE	D	–	S	P	●	●				●					
ZIRCONIUM POTASSIUM FLUORIDE	D	–	S	P	●	●				●					
ZIRCONIUM SULFATE	D	3.22	S	P	●	●				●					
ZIRCONIUM TETRACHLORIDE	D	2.8	S	P	●	●				●					

Source: Akers, C.K., R.J. Pilie, and J.G. Michalovic, 1976, Guidelines of the Use of Chemicals in Removing Hazardous Substance Discharges, prepared under Contract No. 68-03-2093 Exhibit B for National Environmental Research Center, Office of Research and Development, U.S. Environmental Protection Agency, Cincinnati, OH.

Part II

Application of Technical Codes

The information in Part II is from *Toxic Substance Storage Tank Containment Assurance and Safety Program: Guidelines for Application of Technical Codes for Hazardous Materials Storage Tanks,* prepared by Ecology and Environment, Inc. and Whitman, Requardt and Associates for the State of Maryland, Department of Health and Mental Hygiene, under a grant from the U.S. Environmental Protection Agency, September 1983.

1. Introduction

Over the years, technical codes have been developed by independent trade organizations to standardize materials, design, fabrication, and inspection methodologies in a variety of industrial areas. These codes delineate acceptable and desirable practices an industry should follow to attain uniform quality and safety in its operations. The general purpose of such codes is to ensure reasonable protection of life and property, and to adequately limit in-service product deterioration, so as to provide a reasonably long and safe period of usefulness. Adherence to such codes is not mandatory, however, unless they are specially adopted by a regulatory body.

Technical codes are generally developed by consensus of a committee within an independent or industrial organization. Such committees typically are composed of technically qualified representatives of producers, users, and general interest groups. When codes are adopted by a regulatory agency, representatives of that agency often are invited to participate in committee actions to ensure correspondence between the technical code and other regulations in effect in their jurisdiction. Engineers, designers, manufacturers, and inspectors should determine if the applicable state and local regulations differ from the technical codes they are based upon.

The technical codes promulgated by an organization take many forms which are variously called standards, specifications, or recommended practices. <u>Standards</u>, a term often used interchangeably with "code," is an inclusive term that designates the specifications, methods, definitions, classifications, or practices adopted by an organization. Standards are generally regarded as mandatory criteria with which a producer must comply if the product is to bear the organization's seal of approval. <u>Specifications</u> are a precise statement of the requirements to be satisfied by a material or product, and the procedures to ensure adherence to those requirements. The specifications are usually expressed numerically and incorporated directly or by reference in the technical standards. <u>Recommended practices</u> are procedures and specifications which are advisable in nature, not mandatory for gaining the sponsoring organization's approval.

Because the codes may be either more or less restrictive than needed for certain applications, their use is generally voluntary, and the mere existence of technical codes or standards does not preclude anyone from using, purchasing, manufacturing, or marketing products or procedures which do not conform to the standards. Frequently, technical codes will form the basis for regulatory control of a particular class or aspect of industry. In such cases a given technical code forms the standard which a governmental agency will impose through a regulatory process. Commonly the agency will also adopt a program of licensing or inspection to insure that the provisions of those codes adopted through regulation are being met. Although technical codes cannot be construed as being binding in a regulatory or legal sense unless specifically adopted by a regulating agency, they may be a prerequisite to obtaining adequate insurance coverage.

2. Codes Applicable to Storage Tanks

Ideally the application of technical codes should commence at the storage system design stage. Several codes detail the design and fabrication of storage tanks, and these constitute the acceptable industrial standards. These codes are typically applicable to specific tank types, tank contents, construction materials, construction methods, or types of appurtenances. Since there is much variability and overlap among the codes, design and inspection engineers should consult individual codes to determine their specific scope and limitations. Many of the major technical codes applicable to various aspects of storage system design are listed in Table 2-1. This table indicates the title and number of codes, and the types of tanks to which they apply.

2.1 ATMOSPHERIC TANKS

A series of specifications for steel atmospheric storage tanks has been developed by the American Petroleum Institute (API). These include:

- API Spec. 12 B, Bolted Production Tanks;
- API Spec. 12 D, Large Welded Production Tanks; and
- API Spec. 12 F, Small Welded Production Tanks.

Standards for aluminum tanks are covered by ANSI (American National Standards Institute) Standard B 96.1.

Many standards dealing with water handling and storage have been developed by the American Water Works Association (AWWA). Although these tanks are intended for water storage, they are applicable for storage of other liquids as well. Their standards include:

- AWWA D 100-67 - Standard for Steel Tanks, Standpipes, Reservoirs, and Elevated Tanks for Water Storage;

- AWWA D 101-53 - Standard for Inspecting and Repairing Steel Tanks, Standpipes, Reservoirs, and Elevated Tanks for Water Storage; and

- AWWA D 102-64 - Standard for Painting and Repainting Steel Tanks, Standpipes, Reservoirs, and Elevated Tanks for Water Storage.

Underwriters Laboratories (UL) has developed standards which overlap those of API, but include tanks too small for API standards.

Table 2-1

MAJOR TECHNICAL CODES APPLICABLE TO STORAGE SYSTEMS

Organization - Code No.			Title	Applicable Tank Types
API	Spec.	12B	Bolted tanks for storage of production liquids	A
API	Spec.	12D	Field welded tanks for storage of production liquids	A
API	Spec.	12F	Shop welded tanks for storage of production liquids	A
API	RP	12RI	Setting, connecting, maintenance, and operation of lease tanks	A
API	Std.	510	Pressure vessel inspection code	L, H
API	RP	520	Design and installation of pressure-relieving systems in refineries	H
API	RP	521	Pressure relief and depression systems	H
API	Std.	526	Flanged steel safety relief valves	L, H
API	Std.	620	Design and construction of large welded, low-pressure storage tanks	L
API	Std.	650	Welded steel tanks for oil storage	A
API	Publ.	1587	Waste oil round-up	A
API	Publ.	1604	Abandonment or removal of used underground service station tanks	A
API	Bull.	1615	Installation of underground petroleum storage systems	A
API	Publ.	1621	Bulk liquid stock control at retail outlets	A
API	Bull.	1623	Bulk liquid loss control in terminals and depots	A
API	Bull.	1628	Underground spill clean-up manual	A, L, H
API	Std.	2000	Venting atmospheric and low-pressure storage tanks	A, L
API	RP	2001	Fire protection in refineries	A, L, H
API	RP	2003	Protection against ignitions arising out of static, lightning, and stray currents	A, L, H
API	Publ.	2009	Safe practices in gas and electric cutting and welding in refineries, gasoline plants, cycling plants, and petrochemical plants	A, L, H
API	Publ.	2013	Cleaning Mobile tanks in flammable or combustible liquid service	A
API	Publ.	2015	Cleaning petroleum storage tanks	A, L, H
API	Publ.	2015A	A Guide for controlling the lead hazard associated with tank cleaning and entry	A, L, H
API	Publ.	2023	Safe storage and handling of petroleum-derived asphalt products and crude oil residues	A
API	Bull.	2202	Dismantling and disposing of steel from tanks which have contained leaded gasoline	A

Table 2-1 (Cont.)

Organization - Code No.			Title	Applicable Tank Types
API	Publ.	PSD-2207	Preparing tank bottoms for hot work	A, L, H
API	Std.	2510	Design and construction of LPG installations at marine terminals, natural gas plants, refineries, and tank farms	L, H
API	Bull.	2519	Use of internal floating covers and covered floating roofs to reduce evaporation loss.	A
API			Guide for Inspection of Refinery Equipment:	
			• Ch. II - Conditions causing deterioration or failures	A, L, H
			• Ch. III - General preliminary and preparatory work	A, L, H
			• Ch. IV - Inspection tools	A, L, H
			• Ch. V - Preparation of equipment for safe entry and work	A, L, H
			• Ch. VI - Pressure vessels	L, H
			• Ch. XI - Pipes, valves, and fittings	A, L, H
			• Ch. XII - Foundations, structures, and buildings	A, L, H
			• Ch. XIII - Atmospheric and low-pressure storage tanks	A, L
			• Ch. XIV - Electrical systems	A, L, H
			• Ch. XV - Instruments and control equipment	A, L, H
			• Ch. XVI - Pressure relieving devices	A, L, H
			• Ch. XVII - Auxiliary and miscellaneous equipment	A, L, H
			• Appendix - Inspection of welding	A, L, H
API			Guide for Follow-up Inspection of Interior Tank Coatings	A, L, H
NFPA		11	Foam extinguishing systems	A, L, H
NFPA		11A	High expansion foam systems	A, L, H
NFPA		11B	Synthetic foam and combined agent systems	A, L, H
NFPA		12	Carbon Dioxide extinguishing systems	A, L, H
NFPA		12A	Halogenated fire extinguishing agent systems	A, L, H
NFPA		16	Installation of foam-water sprinkler systems and foam-water spray systems	A, L, H
NFPA		17	Dry chemical extinguishing systems	A, L, H
NFPA		30	Code for flammable and combustible liquids	A, L, H
NFPA		43A	Liquid and solid oxidizing materials	A, L, H
NFPA		49	Hazardous chemical data	A, L, H

Table 2-1 (Cont.)

Organization	- Code No.	Title	Applicable Tank Types
NFPA	58	Storage and handling of LPG	L, H
NFPA	59	Storage and handling of LPG at utility gas plants	L, H
NFPA	68	Explosion venting	A, L, H
NFPA	69	Explosion preventing systems	A, L, H
NFPA	70	National electrical code	A, L, H
NFPA	72A	Installation, maintenance, and use of local protective signaling systems	A, L, H
NFPA	72B	Installation, maintenance, and use of auxiliary protective signaling systems	A, L, H
NFPA	72C	Installation, maintenance, and use of remote protective signaling systems	A, L, H
NFPA	72D	Installation, maintenance, and use of proprietary protective signaling systems	A, L, H
NFPA	72E	Automatic fire detectors	A, L, H
NFPA	77	Recommended practice on static electricity	A, L, H
NFPA	78	Lightning protection code	A, L, H
NFPA	231	General indoor storage	A
NFPA	231A	General outdoor storage	A
NFPA	321	Classification of flammable and combustible liquids	A, L, H
NFPA	325M	Fire hazard properties of flammable liquids	A, L, H
NFPA	327	Cleaning small tanks and containers	A, L, H
NFPA	329	Underground leakage of flammable and combustible liquids	A, L, H
NFPA	419M	Code for explosive materials	A, L, H
NFPA	495	Identification of fire hazards of materials	A, L, H
NFPA	1221	Installation, maintenance, and use of public fire service communications	A, L, H
NFPA		Fire Protection Guide on Hazardous Materials	
ASME		Boiler and Pressure Vessel Code: o Section II - Materials specifications o Section V - Nondestructive examination o Section VIII - Pressure vessels o Section X - FRP Pressure vessels	L, H
AWWA	D100-67	Standard for steel tanks, standpipes, reservoirs, and elevated tanks for water storage	A
AWWA	D101-53	Standard for inspecting and repairing steel tanks, standpipes, reservoirs, and elevated tanks for water storage	A

Table 2-1 (Cont.)

Organization	- Code No.	Title	Applicable Tank Types
AWWA	D102-64	Standard for painting and repainting steel tanks, standpipes, reservoirs, and elevated tanks for water storage	A
ACI		Guide for Protection of Concrete Against Chemical Attack by Means of Coatings and Other Corrosion-Resistant Materials	A, L, H
ACI		Manual of Concrete Practices	A, L, H
ACI	344	Design and construction of circular prestressed concrete structures	A, L, H
AIA		Fire Prevention Code	A, L, H
NACE	RP-01-69	Control of external corrosion on underground or submerged metallic piping systems	A, L, H
NACE	No. 1	Surface preparation for tank linings	A, L, H
NACE	No. 2	Surface preparation for some tank linings and heavy maintenance	A, L, H
NACE	No. 3	Surface preparation for maintenance	A, L, H
NACE	No. 4	Surface prepartion for very light maintenance	A, L, H
NACE	RP-03-72	Method for lining lease production tanks with coal tar epoxy	A, L, H
SSPC	5-63	White metal blast	A, L, H
SSPC	10-63	Near-white metal blast	A, L, H
SSPC	6-63	Commercial blast	A, L, H
SSPC	7-63	Brush off blast	A, L, H

LEGEND:

Organization:
API = American Petroleum Institute
NFPA = National Fire Protection Association
ASME = American Society of Mechanical Engineers
AWWA = American Water Works Association
ACI = American Concrete Institute
NACE = National Association of Corrosion Engineers
SSPC = Steel Structures Painting Council

Code Number:
A numerical designation assigned to a code, etc., by the promulgating organization.
Spec = specification
RP = recommended practice
Std = standard
Publ. = publication
Bull. = bulletin

Applicable Tank Types:
A = Atmospheric
L = Low Pressure
H = High Pressure

Source: Ecology and Environment, Inc., 1983.

UL standards do not provide as much detail as API standards, and therefore put more responsibility on the designer to obtain guidance from other professional reference standards and guidelines. Two UL standards applicable to hazardous materials storage are:

- UL 58 - Steel Underground Tanks for Flammable and Combustible Liquids; and

- UL 142 - Steel Aboveground Tanks for Flammable and Combustible Liquids.

Other standards related to atmospheric tanks include:

- ANSI B31.3 - Petroleum Refinery Piping Code;

- API Standard 650 - Welded Steel Tanks for Oil Storage; and

- API Standard 2000 - Venting Atmospheric and Low-Pressure Storage Tanks.

2.2 LOW-PRESSURE TANKS

Vertical, cylindrical tanks with domed roofs which operate at pressures slightly above atmospheric pressure can be built according to API Standard 650. However, for tanks with more substantial operating pressures, up to 15 psig, API Standard 620, Recommended Rules for Design and Construction of Large, Welded, Low-Pressure Storage Tanks, establishes the proper design criteria. Section VIII of the American Society of Mechanical Engineers (ASME) Boiler and Pressure Vessel Code may provide useful design guidelines, although they are not directly applicable to tanks with operating pressures below 15 psig.

2.3 HIGH-PRESSURE TANKS

In general terms, a pressure vessel is a closed container of limited length which is subject to pressures above one or two pounds per square inch. For purposes of this discussion, a high-pressure tank or vessel is one with an operating pressure greater than 15 psig.

The most comprehensive standards for the design, fabrication, and inspection of high-pressure tanks is found in the ASME Boiler and Pressure Vessel Code. This is an 11-section publication that is issued every three years. Most states have passed regulations which make all or certain parts of the ASME Code legal requirements, and have inspectors to enforce provisions of the Code. Sections of the ASME Code applicable to the design and inspection of high-pressure storage tanks are:

- Section II - Material Specifications;

- Section V - Nondestructive Examination;

- Section VIII - Pressure Vessels, Division 1;
  Pressure Vessels, Division 2 - Alternative Rules;

- Section IX - Welding Qualifications; and

- Section X - Fiberglass-Reinforced Plastic Pressure Vessels.

Other codes pertinent to high-pressure storage systems have been developed by the American Petroleum Institute and the National Fire Protection Association. These include:

- API 510 - Pressure Vessel Inspection Code;
- API - Guide for the Inspection of Refinery Equipment, Chapter VI, Pressure Vessels;
- API Standard 2510 - Design and Construction of LP Gas Installations at Marine and Pipeline Terminals, Natural Gas Processing Plants, Refineries, and Tank Farms;
- NFPA 58 - Liquefied Petroleum Gases, Storage and Handling;
- NFPA 59A - Liquefied Natural Gas, Storage and Handling; and
- NFPA 43C - Oxidizing Materials, Gaseous, Storage.

2.4 STORAGE TANK CONSTRUCTION MATERIALS

Selection of the appropriate tank materials is a critical element in the design of a hazardous materials storage system. Although literally thousands of codes exist for all types of construction materials, the scope of those standards and specifications does not necessarily address their application to hazardous materials storage. Therefore, the designer is advised to consult the codes, such as those identified in Sections 2.1 through 2.3, most applicable to the type of storage under consideration. The codes will generally identify, or give references to, pertinent materials standards. As an example, the base materials specifications in Parts A and B of the ASME Boiler and Pressure Vessel Code, Section II, are similar or identical to those of the American Society for Testing and Materials (ASTM). The design engineer may then refer to the cited ASTM standard for further details.

ASTM currently lists over 6,700 standards in its 66-volume, 16-section 1983 Annual Book of ASTM Standards (prior to 1983, the ASTM standards were published in 48 parts). Those sections containing material and procedural standards applicable to hazardous materials storage system design are:

- Section 1 - Iron and Steel Products;
- Section 2 - Nonferrous Metal Products;
- Section 3 - Metal Test Methods and Analytical Procedures;
- Section 6 - Paints, Related Coatings, and Aromatics;
- Section 8 - Plastics;
- Section 9 - Rubber;
- Section 11 - Water and Environmental Technology;

- Section 14 - General Methods and Instrumentation; and
- Section 15 - General Products, Chemical Specialties, and End Use Products.

A complete listing of ASTM standards and a cross reference between 1983 volumes and previous year part numbers may be obtained from:

American Society for Testing and Materials
1916 Race Street
Philadelphia, PA 19103
(215) 299-5462

Another source of design specifications frequently cited or incorporated in techincal codes is the American National Standards Institute (ANSI). This organization coordinates America's voluntary standards system, maintaining an inventory of over 11,000 American standards. ANSI also maintains the 6,000 standards promulgated by the International Organization for Standardization (ISO) and the International Electrotechnical Commission (IEC). Standards of the 71 national standardizing bodies of other countries cooperating within the ISO are stocked or obtainable by ANSI. Listings of available standards may be obtained from:

American National Standards Institute, Inc.
1430 Broadway
New York, New York 10018
(212) 354-3300

Further information about materials standards and specifications may be obtained from a variety of industry-specific organizations. These include:

American Petroleum Institute
2101 L Street, NW
Washington, DC 20037
(202) 457-7160

American Society of Mechanical Engineers
345 E. 47th Street
New York, NY 10017
(212) 705-7722

American Iron and Steel Institute
1000 16th Street, NW
Washington, DC 20036
(202) 452-7100

Steel Structures Painting Council
4400 5th Avenue
Pittsburgh, PA 15213
(412) 578-3327

National Association of Corrosion Engineers
1440 South Creek
Houston, TX 77084
(713) 492-0535

American Welding Society
2501 NW Seventh Street
Miami, FL 33125
(305) 642-7090

American Institute of Chemical Engineers
345 E. 45th Street
New York, NY 10017
(212) 705-7338

Steel Tank Institute
666 Dundee Road, Suite 705
Northbrook, IL 60062
(312) 498-1980

Society of the Plastics Industry
355 Lexington Avenue
New York, NY 10017
(212) 573-9400

Chemical Manufacturers Association
2501 M Street, NW
Washington, DC 20037
(202) 887-1100

National Fire Protection Association
Batterymarch Park
Quincy, MA 02269
(617) 328-9230

Underwriters Laboratory, Inc.
333 Pfingston Road
Northbrook, IL 60062
(312) 272-8800

American Society for Metals
Metals Park, OH 44073
(216) 338-5151

It should be noted that materials specifications typically address general issues such as metal strength and ductility. Material selection data for more specific considerations, such as chemical compatibility, should be determined on a case-by-case basis according to the best professional judgement.

3. Use and Interpretation of Technical Codes

As previously indicated, technical codes are developed to delineate acceptable and desirable practices an industry should follow to attain uniform quality and safety in its products and operations. The codes are generally intended for use by design engineers and technically qualified inspectors, and consequently are often written in highly technical language. To most laymen, the sheer number of technical codes and their voluminous technical detail can make it impractical to gain a thorough understanding of the codes. However some guidance can be provided to simplify use of the codes.

The first step in code usage is to select the code or codes, such as those identified in Sections 2.1 to 2.4, most applicable to the subject under consideration. These codes are usually identified in their titles as "standards," "specifications," "recommended practices," or by other similar descriptives. It is necessary to understand the distinction between these terms, as discussed in Section 1, to understand the context of the various codes.

Upon identification of the codes to be studied, it is essential that the user thoroughly read the foreword, preface, and any other introductory remarks about the code. These sections usually provide valuable information about the basis and derivation of the codes, their purpose and intent, limits of application, disclaimers, dates of revision, etc., and clearly define the context within which the code should be interpreted.

The next step is to review the table of contents. Besides identifying the subject matters within the code, this will familiarize the reader with the format, nomenclature, and content subdivisions. This will be of aid later for cross-referencing between subsections. Note the presence and contents of any appendices. Often the appendices contain additional standards or recommendations which would not otherwise be identified in a catalog of technical code titles, and they may also include definitions or other information which would be beneficial to read before reading the actual technical standards.

At this point, the reader should be able to identify the section and subsections of the code dealing with subjects of interest. However, it is still important to read the introductory paragraphs of the individual sections to determine the scope and limitations of those sections.

A consistent format is used throughout most codes to designate each of the sections, subsections, or paragraphs within the code. These designations may be in the form of letters, numbers, or a combination of both, arranged in sequential order. Each subsection is identified by the section designator, followed by a hyphen or period and a number in sequence (e.g., A-1, A-2; 2.1, 2.2, 2.3, etc.). This facilitates reference to particular paragraphs or subsections without the need for extensive page references.

Once the proper code has been identified, and all applicable introductory and background information has been read, use of the code is simply a matter of reading the standards, referring to the proper cross references, and correctly interpreting what is written.

The use of four technical codes which provide the most information in the design, fabrication, and inspection of hazardous materials storage tanks are described in the following sections.

3.1 API STANDARDS AND SPECIFICATIONS

The American Petroleum Institute (API) has published hundreds of documents relevant to the production and refining of oil and petroleum products. Although originally developed for the petroleum industry, these standards can be applicable to most hazardous materials storage issues, provided that such specific issues as chemical compatibility are properly addressed and incorporated by the designer.

API codes are usually referred to as either "standards" or "specifications." According to API interpretations, these terms generally are identical in meaning. The use of one term or the other depends on the API division which developed the code. The API Production Department's codes are referred to as "specifications," and are generally applicable to atmospheric storage tanks for production fluids. "Standards" are promulgated by the API Refining Department, and are applicable to low- and high-pressure tanks.

Both API standards and specifications are designated by number, with subsection indicated sequentially by decimal places. Cross-references to standards within the same code document are designated by the paragraph or subsection number only. References to other API codes are made by title and number of the other reference. Material specifications are usually handled by referencing (by number and title) the appropriate ASTM or ANSI standards, unless API data is incorporated directly into the code.

In addition to its standards and specifications, API also publishes various recommended practices, bulletins, and publications. These documents do not set forth specific criteria or minimum standards that tank designs must meet before receiving API approval. However, they are useful as sources of further information about problems, hazards, or other practices related to the standards and specifications.

3.2 NATIONAL FIRE CODES

The _National Fire Codes_ are an annual, 16-volume publication of the _National Fire Protection_ Association (NFPA). These volumes are compilations of various individual codes, standards, recommended practices, manuals, guides, and model laws prepared by the NFPA. Only those documents which have been adopted by NFPA are included in the National Fire Codes.

Volumes 1 through 12 of the _National Fire Codes_ contain NFPA codes and standards judged _suitable for legal adoption_ and enforcement by government agencies. Volumes 13 through 16 contain recommended practices, manuals, and guides identifying good engineering practices. These volumes also include model laws and enabling acts which may be of help to regulatory bodies. Many of the NFPA documents have been approved as standards by ANSI.

Most of the documents in the National Fire Codes are available as separate pamphlets. A complete listing of these is found in the inside front and back covers of each volume of the National Fire Codes. This listing identifies the document's NFPA Code Number and the National Fire Codes volume number in which the listed title of interest may be found. This listing is suggested as the starting point for locating applicable NFPA standards.

Use of the NFPA codes is similar to the use of API codes. Each code is divided into chapters covering different elements of the code. The chapters are then divided into subsections indicated by the chapter number, a hyphen, and sequential subsection numbers. Paragraphs within each subsection are further indicated by sequential decimal places following the subsection number. For example, NFPA 30, Flammable and Combustible Liquids Code, contains:

CHAPTER 2 TANK STORAGE

2-1 Design and Construction of Tanks
2-1.1 Materials
2-1.2 Fabrication
2-1.2.1 (Paragraph 1)
2-1.2.2 (Paragraph 2)

Pages within the code are indicated by sequential page numbers preceded by the code number (e.g., page 30-17). References to other portions of the same code are indicated by only the section, subsection, or paragraph numbers. References to other codes refer to the title and number of the other code.

3.3 ASTM STANDARDS

The 1983 Annual Book of ASTM Standards contains over 6,700 current standards, divided into 16 sections of related subject areas. These sections are published in 66 volumes which further subdivide standards for related items. (Prior to 1983, the ASTM Standards were published in 48 parts.) Location of applicable ASTM standards is facilitated by use of the index, Section 00 (Part 48 in pre-1983 editions).

The index provides a listing of ASTM Standards by subject and by ASTM's alphanumeric code listing. The main subject entries refer to subjects taken directly from the scope of the standard, or they may highlight similar standards. In either case, the subject listing will be followed by the standard number in parentheses, and the number of the section where it may be found. Cross references within the index guide the user from related terms to the listed subject entries. These cross references are designated as "See" or "See Also" (or Sa) written in italics.

Each adopted or tentative ASTM standard has an individual serial designation listed in the alphanumeric index. This consists of a capital letter, a serial number of one to four digits, a hyphen, and a two-digit number indicating the year of adoption or latest revision. The letter at the beginning of the serial number indicates the general classification of the standard:

A = Ferrous Metals,
B = Nonferrous Metals;

C = Cementitious, Ceramic, Concrete, and Masonry Materials;
D = Miscellaneous Material;
E = Miscellaneous Subjects;
F = Materials for Specific Applications;
G = Corrosion, Deterioration, and Degradation of Materials; and
ES = Emergency Standards.

The serial number may be followed by a lower case letter which denotes a revision occurring in the same year as the year of adoption or latest revision. A capital letter "T" at the end of the standard is used to denote a tentative standard. The letter "M" after the serial number designates a standard in metric units.

The standard number may be followed by a number in parentheses, which indicates the year the standard was last reapproved. As an example, A236-69a (1974) refers to a specification for carbon steel forgings for railway use, first adopted in 1969, indicated by the "69" after the hyphen; then revised in 1969, indicated by the letter "a"; and reapproved in 1974, indicated by that date in parentheses.

The title is given following the standard's serial designation. A boldface number following the title indicates the volume in which the standard may be found.

The ASTM standards are cited frequently as design criteria by such organizations as API or ASME. When cited, the complete designation of standard number and title is usally given. The alphanumeric index may be used to locate the specific reference.

3.4 ASME BOILER AND PRESSURE VESSEL CODE

The American Society of Mechanical Engineers wrote the ASME Boiler and Pressure Vessel Code to provide rules for the design, fabrication, and inspection of boilers and pressure vessels. The ASME Code is written by a large Boiler and Pressure Vessel Committee and several subcommittees composed of engineers appointed by ASME. The Code Committee meets regularly to review the Code and to consider requests for its interpretation or revision. Interpretations and special provisions adopted by the committee are published in "Mechanical Engineering," and in a supplement to the ASME Code entitled "Code Cases."

By itself, the ASME Code has no legal status. However, most states have adopted at least parts of the Code as their legal requirements, and employ inspectors to enforce provisions of the Code. A new edition of the ASME Code is issued every three years, with the latest edition scheduled for 1983. Addenda to the Code are published semiannually between editions. Although ASME considers any edition of the Code to be adequate, certain issues of the Code may be specified as legal requirements by some regulatory bodies.

The chief inspectors of all states and municipalities which have adopted the code form the National Board of Boiler and Pressure Vessel Inspectors. The Board promotes uniform enforcement of Code rules by examining and commissioning inspectors. Qualified inspectors employed by a state, municipality, or insurance company may inspect a pressure vessel and permit it to be stamped "ASME-N.B." (The "N.B." stands for "National Board"). A vessel bearing the "N.B." stamp and registered with the National Board generally may be used and sold in any state. Inspectors employed by a vessel user may only use the ASME stamp, which somewhat limits the use of those vessels.

As indicated in Section 2.3, the ASME Code is published in 11 Sections. The user of the code should begin with the section title deemed most likely to cover the subject of interest. For storage applications, "Section VIII, Pressure Vessels, Division 1," and "Pressure Vessels, Division 2, Alternative Rules" would be the most likely starting point. Division 1 is the Pressure Vessel Code as it has existed in the past. Division 2 was developed as a means of permitting higher design stresses while ensuring at least as great a degree of safety as Division 1.

Section VIII, Division 1 is divided into three subsections: A, General Requirements; B, Fabrication Methods; and C, Materials of Construction. These subsections are further divided into a total of 13 Parts, which are designated by two letters, the first of which is "U" for "unfired." These are followed by two groups of appendices. The mandatory appendices, indicated by Roman numerals, detail procedures frequently referred to in the ASME Code. The non-mandatory appendices, designated by capital letters, provide information and suggested procedures which are not essential, but may be helpful to Code users. Of particular help to someone new to the Code are Appendices III and L. Appendix III defines many of the terms used in the Code, and is a good starting point for a first-time user. Appendix L provides sample problems illustrating application of Code formulas and rules, and is useful to the novice designer.

Section VIII, Division 2 is more sophisticated than Division 1, and requires a more technically competent engineer to use it. Division 1 employs a safety factor of about 4 in its standards, but ignores many secondary stresses which act in the vessel. Division 2 allows higher stresses, but requires thorough stress analysis and closer quality control of materials and fabrication.

Because of the extensive nature of the ASME Code, the user is advised to pay particular attention to background information, tables of contents, and indices (found at the end of each section) before using the Code. Although the style of designating parts, subsections, or articles may vary from section to section, each designation is unique, and cross-referencing within a section is accomplished rather easily. References to other sections of the Code will include the title and section number.

4. Conclusions and Recommendations

As discussed in Section 3, technical codes are quite complex, and often difficult to understand. They provide a tremendous amount of technical detail which may easily overwhelm the user if the codes are not approached systematically. This requires selection of the appropriate code with which to begin the search for information.

As illustrated by Table 2-1, there are many technical codes applicable to specific aspects of hazardous materials storage systems. Selection of the inappropriate code may result in failure to determine standards specifically applicable to the problem at hand, and it may lead to an unnecessarily laborious search for the correct information. Therefore, selection of the code which appears to offer the broadest coverage of the subject of interest will either provide the information sought or direct the user to other pertinent sources.

It is important to consult those codes promulgated by organizations most involved with the type of materials to be used. For example, Section 2.1 identified standards of the American Petroleum Institute and the American Water Works Association (AWWA) which are applicable to atmospheric storage tanks. Because of API's extensive experience with the petroleum and petrochemical industry, its codes could be expected to provide more information relevant to hazardous materials storage, whereas the AWWA standards would have much more limited application. Therefore, because of their relevance and broad scope, it is recommended that the designer searching for standards applicable to specific types of storage tanks begin the search with the following codes:

- API Specification 12D and API Standards 650 for atmospheric tanks;
- API Standard 620 for low-pressure tanks; and
- ASME Boiler and Pressure Vessel Code, Section VIII for high-pressure tanks.

These codes will in turn cross reference the appropriate codes for materials, appurtenances, etc.

Although it may seem obvious, elementary, and therefore, unnecessary, the foreword, table of contents, and introductory paragraphs must be read if the codes are to be properly understood. To the casual user reading the code for informational purposes, the introductory paragraphs describe the framework within which the code is applicable. To the designer, the introductory material may identify the scope and application of the code and other sources of valuable information. Because codes are revised regularly, it is also important that the user consult the latest applicable edition.

Technical codes are written by and for technical experts within a particular field. Therefore, the technical language may be unfamiliar to some users. Interpretation of the code, and formulation of judgements based on those interpretations, should always be left to individuals technically qualified to make such interpretations.